Biomedical Signal Processing

Volume II
Compression and Automatic Recognition

Author

Arnon Cohen, Ph.D.
Associate Professor
Departments of Biomedical Engineering and
Electrical and Computer Engineering
Head
Center for Biomedical Engineering
Ben Gurion University
Beer Sheva, Israel

CRC Press
Taylor & Francis Group
Boca Raton London New York

CRC Press is an imprint of the
Taylor & Francis Group, an **informa** business

CRC Press
Taylor & Francis Group
6000 Broken Sound Parkway NW, Suite 300
Boca Raton, FL 33487-2742

Reissued 2019 by CRC Press

A Library of Congress record exists under LC control number:

ISBN 13: 978-0-367-25965-5 (hbk)
ISBN 13: 978-0-367-25968-6 (pbk)
ISBN 13: 978-0-429-29080-0 (ebk)

Visit the Taylor & Francis Web site at http://www.taylorandfrancis.com and the CRC Press Web site at http://www.crcpress.com

Dedicated to
my mother Rachel
my wife Yama
and my sons Boaz, Gilead, and Nadav

PREFACE

Biomedical signal processing is of prime importance not only to the physiological researcher but also to the clinician, the engineer, and the computer scientists who are required to interpret the signal and to design systems and algorithms for its manipulations.

The biomedical signal is, first of all, a signal. As such, its processing and analysis are covered by the numerous books and journals on general signal processing. Biomedical signals, however, possess many special properties and unique problems that render the need for special treatment.

Most of the material dealing with biomedical signal processing methods has been widely scattered in various scientific, technological, and physiological journals and in conference proceedings. Consequently, it is a rather difficult and time-consuming task, particularly to a newcomer to this field, to extract the subject matter from the scattered information.

This book was not meant to be another text or reference on general signal processing. It is intended to provide material of interest to engineers and scientists who wish to apply modern signal processing techniques to the analysis of biomedical signals. It is assumed the reader is familiar with the fundamentals of signals and systems analysis as well as the fundamentals of biological systems. Two chapters on basic digital and random signal processing have been included. These serve only as a summary of the material required as background for other material covered in the book.

The presentation of the material in the book follows the flow of events of the general signal processing system. After the signal has been acquired, some manipulations are applied in order to enhance the relevant information present in the signal. Simple, optimal, and adaptive filtering are examples of such manipulations. The detection of wavelets is of importance in biomedical signals; they can be detected from the enhanced signal by several methods. The signal very often contains redundancies. When effective storing, transmission, or automatic classification are required, these redundancies have to be extracted. The signal is then subjected to data reduction algorithms that allow the effective representation in terms of features. Methods for data reduction and features extraction are discussed. Finally, the topic of automatic classification is dealt with, in both the decision theoretic and the syntactic approaches.

The emphasis in this book has been placed on modern processing methods, some of which have been only slightly applied to biomedical data. The material is organized such that a method is presented and discussed, and examples of its application to biomedical signals are given. Rapid developments in digital hardware and in signal processing algorithms open new possibilities for the applications of sophisticated signal processing methods to biomedicine. Solutions that were cost prohibitive beforehand or impractical because of the lack of appropriate algorithm become available. In such a dynamic environment, the biomedical signal processing practitioner requires a book such as this one.

The author wishes to acknowledge the help received from many students and colleagues during the preparation of this book.

Arnon Cohen

THE AUTHOR

Arnon Cohen, Ph.D., is an Associate Professor of Electrical and Bio-Medical Engineering at the Ben-Gurion University, Beer-Sheva, Israel, and the head of the Center for Bio-Medical Engineering.

Dr. Cohen received his B.Sc. and M.Sc. degrees in Electrical Engineering in 1964 and 1966 respectively, from the Israel Institute of Technology (Technion) in Haifa. He received his Ph.D. in 1970 from the Department of Electrical Engineering (Bio-Medical Engineering Program), Carnegie Mellon University in Pittsburgh.

Prior to joining the Ben-Gurion University in 1972, he was an Assistant and later an Associate Professor of Electrical and Bio-Medical Engineering at the Connecticut State University, Storrs.

During 1976-1977, Dr. Cohen was appointed a Visiting Professor at the Electrical Engineering department of Colorado State University, Fort Collins.

Dr. Cohen has been the recipient of research grants from various foundations and corporations in the U.S. and Israel. He has been a consultant with various industrial companies in the U.S. and Israel. Recently, he has founded and became the president of MEDI-SOFT, a consulting firm in the field of bio-medical signal processing. Dr. Cohen's current research interests are signal and speech processing with applications to biomedicine.

TABLE OF CONTENTS

Volume I

Chapter 4
Digital Signal Processing

Chapter 5
Finite Time Averaging

Chapter 6
Frequency Domain Analysis

TABLE OF CONTENTS

Volume II

Chapter 1

WAVELET DETECTION

I. INTRODUCTION

A common problem in biomedical signal processing is that of detecting the presence of a wavelet in a noisy signal. A wavelet is considered here as any real valued function of time, possessing some structure. The approximate shape of the wavelet, expected to be present in the noisy signal, may be known and the requirements are to estimate its time of occurrence and its exact shape. The problem, of course, is not restricted to biomedical signals; it appears, for example, in communications, radar, sonar, or seismological signals. Two main approaches are used to solve the problem. The first uses structural features of the wavelet and the second uses template matching techniques.

The approximate knowledge of the wavelet's shape can be used to select typical structural features. These features are random variables with some empirically determined probability density functions. The noisy signal is then continuously searched for these features.

The search can be conducted in various ways. Sophisticated algorithms that can be used to recognize a wavelet, once its features have been extracted, are discussed in Chapters 3 and 4. These algorithms are based on the decision-theoretic or syntactic approaches to pattern recognition. In this chapter we shall consider much simpler techniques, which are usually implemented by relatively inexpensive, dedicated hardware. The algorithms are primarily heuristic and are usually specific to a given wavelet (say the QRS complex in the ECG).

A more general approach is the one where the approximate knowledge on the wavelet is used to generate a template, which is some average wavelet determined by the *a priori* knowledge on the signal. The signal is then continuously matched with the template by means of correlation, matched filtering, or other pattern recognition techniques. The method is general in the sense that one can use the same algorithm for the detection of any wavelet by just replacing the template. The choice between these two approaches depends, among others, on the wavelet, the signal-to-noise ratio, available time of computation, and cost.

Wavelet detection is required in many biomedical signal processing applications. Detection is required for monitoring, alarm systems, and as an initial part of automatic classification. Detection and classification of aperiodic EEG waveshapes[1] in the time domain are used for a variety of clinical applications, such as the treatment of epilepsy or sleep state analysis.[2] The EEG signal contains several aperiodic wavelets which are of clinical interest. These are, for example, spikes, K-complexes, or sleep spindles. Several algorithms for the detection of the various EEG wavelets using both the specific and general methods were reported.[3-10] Evoked response wavelets,[11] as well as multispike train recordings,[12,13] were also detected by such methods. The problem of detecting and classifying wavelets also exists in electrocardiography, for QRS complex detection,[15] fetal heart rate determination,[16] P-wave detection,[17] in phonocardiography for detection and classification of heart sounds,[18] in breath signal analysis,[19] and many more.

In this chapter we shall discuss the problems associated with wavelet detection. First the problem is mathematically formulated. Detection algorithm for time invariant nonoverlapping wavelets is discussed followed by a discussion on an adaptive algorithm. Problems of wavelet alignment, detecting overlapping wavelets and normalization of wavelets by time warping, are also discussed.

II. DETECTION BY STRUCTURAL FEATURES

A. Simple Structural Algorithms

In this section we shall discuss the detection of wavelets in noisy signals by means of simple structural analysis. We assume a class of wavelets, $S_i(t)$, $i = 1,2, \ldots$, all having some common structure. As an example consider the QRS complex in the ECG signal (see Appendix A). The QRS complexes, even consecutive ones, are not identical, but possess common structure of typical Q, R, and S waves. We shall consider wavelets to appear together with additive noise, such that the observation signal x(t) is given by:

$$x(t) = m(t)(\tilde{S}(t) + \tilde{n}(t)) \tag{1.1A}$$

where

$$\tilde{S}(t) = \sum_{i=0}^{\infty} \tilde{G}_i S_i(t - t_i) \tag{1.1B}$$

with

$$S_i(\tau) = 0 \text{ for } \tau < 0$$

$$\tau \geqslant T_i \leqslant T \tag{1.1C}$$

and

$$\frac{1}{T_i}\int_{t_i}^{t_i+T_i} S_i^2(t - t_i)dt = 1 \text{ for all } i \tag{1.1D}$$

In general, the wavelets $S_i(t)$ are stochastic and our goal is to determine the expectation $E\{S_i(t)\}$. In some cases the wavelets are deterministic and we are interested in the exact shape of each one (for example, consider the problem of single evoked potential estimation). Sometimes it will be sufficient to get the mean of several wavelets, say

$$\bar{S}(\tau) = \frac{1}{I}\sum_{i=1}^{I} S_i(\tau)$$

as is the case in average evoked potential analysis.

In Equation 1.1, $\tilde{n}(t)$ is an additive noise generated by other biological signals, which are of no interest for the application at hand, or by the measurement system. The multiplicative noise, m(t), often arises in biological signals. This may be, for example, a modulating noise due to breathing while monitoring the ECG signal.

Equation 1.1 can be written with no direct multiplicative noise by

$$x(t) = S(t) + n(t) \tag{1.2A}$$

where

$$S(t) = \sum_{i=0}^{\infty} m(t)\tilde{G}_i S_i(t - t_i) = \sum_{i=0}^{\infty} G_i(t) S_i(t - t_i) \tag{1.2B}$$

and

$$n(t) = m(t)\tilde{n}(t) \tag{1.2C}$$

where the gains, G_i's, are now functions of time and the noise process, n(t), is nonstationary. The signal, S(t), consists of a series of wavelets S_i, each with gain G_i, appearing at times t_i. Each wavelet is finite in duration and its energy is normalized to 1. We shall also assume, for the time being, that the wavelets do not overlap, namely,

$$T \leqslant (t_{i+1} - t_i) \text{ for all } i \tag{1.3}$$

An *a priori* knowledge on the wavelets, $S_i(t)$, is given in terms of the estimate $\hat{S}(t)$, which is termed "the template". The template $\hat{S}(t)$ is zero outside the range $0 \leqslant t \leqslant T_s \leqslant T$.

The problem of wavelet detection can now be formulated as follows. Given the initial template, $\hat{S}(t)$, estimate the occurrence times, t_i, the exact shape, $S_i(t)$, and the gains, G_i. This is, of course, the general problem. For some applications it may be sufficient just tc determine whether a wavelet was present in a given time window.

The sampled version of Equation 1.2 is given by:

$$x(k) = \sum_{i=0}^{\infty} G_i S_i(k - \tilde{k}_i) + n(k) \tag{1.4A}$$

with

$$S_i(j) = 0 \text{ for } j < 0 \text{ and } j \geqslant M - 1 \tag{1.4B}$$

and

$$\frac{1}{M} \sum_{k=\tilde{k}_i}^{\tilde{k}_i+M-1} S_i^2(k - \tilde{k}_i) = 1 \tag{1.4C}$$

where the sampling interval was assumed to be one without the loss of generality. (If sampling interval is important just replace k and M by $\Delta t \cdot k$ and ΔtM, etc.)

Sophisticated algorithms are available to detect the presence of a wavelet by analyzing its structure. These syntactic methods are discussed in Chapter 4. Several methods exist for shape analysis of waveforms. General discriptors like the Fourier discriptors,[20] polygonal approximations,[21] and others are used mainly for the analysis of two-dimensional pictures, but also for one-dimensional signals. Most often, however, simpler methods are used, specifically designed to the application at hand. The advantage of these methods is their simplicity and the ability to implement it on relatively inexpensive, dedicated hardware. The main disadvantage, however, is the fact that each method is specific to a given wavelet and cannot be generally applied. These schemes are usually rigid and do not lend themselves to adaptation. It is mainly applied to QRS detection[22] and to the detection of wavelets in the EEG.[23] Since methods of the type discussed here depend on the wavelet, the best way to describe it is by an example.

Example 1.1 — QRS complex detection

The detection of the QRS complex in the ECG is required for monitoring and analysis. The QRS complex is the most distinguished part of the PQRST wavelet of the ECG (see Figure 1 and Appendix A). Two main problems prevent the simple detection of the R wave by threshold techniques. First, baseline shifts due to movements of electrodes may place

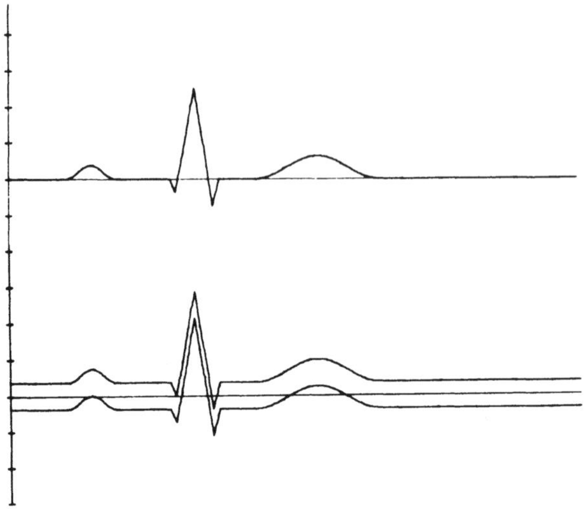

FIGURE 1. Contour limits for QRS complex detection: upper trace, synthesized ECG signal; lower trace, contour lines.

large positive or negative bias in the signal. Second, sometimes line frequency noise may interfere. To overcome these problems, we shall use the sequence of first difference of the observed ECG signal (Equation 1.4). Assume the line frequency is 60 Hz, and we sample the ECG at the rate of 1800 samples per second.

Consider the first difference:

$$\tilde{x}(k) = x(k) - x(k - 30) \tag{1.5}$$

Note that the difference is 30/1800 = 1/60 sec, namely, one period of the line frequency. The first difference is thus synchronized to the line interferences in such a way that these interferences do not appear in $\tilde{x}(k)$ (see discussion on seasonal time series in Chapter 7, Volume I). Note also that base line shifts, which are usually much slower than the power line interferences, will also be eliminated by Equation 1.5.

The wavelet present in the signal, x(k), will now be transformed to

$$\tilde{S}_i(k) = S_i(k) - S_i(k - 30) \tag{1.6}$$

Equation 1.6 may increase the sensitivity of the algorithm to noise, since it is analogous to differentiation.[24] Since the difference operator has removed most of the base line shift, we can now apply threshold techniques. Consider the following threshold[25] procedure. Let

$$\tilde{X}_{MIN} = \underset{k}{\text{Min}}\ \tilde{x}(k)$$

$$\tilde{X}_{MAX} = \underset{k}{\text{Max}}\ \tilde{x}(k) \tag{1.7}$$

The threshold THR is then given by

$$\text{THR} = \tilde{X}_{MIN} + 1/6(\tilde{X}_{MAX} - \tilde{X}_{MIN}) \tag{1.8}$$

The presence of *ith* QRS wavelet is determined at k_i, the *ith* time that the threshold is being crossed:

$$\tilde{x}(k) \geqslant THR \tag{1.9}$$

This algorithm has not used all the structural characteristics of the QRS. The R wave consists of an upslope of typical slope and duration[26] followed immediately by a downslope with characteristic slope and duration. Condition 1.9 can be considered as an hypothesis for QRS complex. The hypothesis can be accepted if the upslope and downslope in the neighborhood of k_i meet the R wave specifications.

The accuracy of the QRS detector is important, especially when high frequency ECG is of interest (see Appendix A). The inaccuracies in the QRS time detection, known as *jitters*, in simple threshold detection systems were discussed by Uijen and co-workers.[30]

Many other algorithms for QRS detection have been suggested.[27-29] A well-known algorithm is the *amplitude zone time epoch coding* (AZTEC).[38] This algorithm has been developed for real time ECG analysis and compression. AZTEC analyzes the ECG waveform and extracts two basic structural features: plateaus and slopes. A plateau is given in terms of its amplitude and duration and a slope by its final elevation and duration. Another algorithm, the *coordinate reduction time encoding system* (CORTES),[39] has been suggested which is an improvement to the AZTEC.

B. Contour Limiting

The methods discussed in the previous section suffer from the fact that they are "tailored" to a specific wavelet. The method of contour limiting[31,32] is more flexible in the sense that it can easily be adapted to various wavelets. The contour limiting uses a template, which is some typical wavelet. The template can be constructed from some *a priori* knowledge about the wavelet or by averaging given wavelets that were detected and aligned manually. Knowledge about the expected variations of each sample of wavelet is also required. From this knowledge upper and lower limits are constructed. Consider the observation vector x(k) of Equation 1.4. The upper limit $S^+(k)$ and lower limit $S^-(k)$ are given by

$$\begin{aligned} S^+(k) &= \hat{G}\hat{S}(k) + L^+(k) \\ k &= 1,2,\ldots,M \end{aligned} \tag{1.10A}$$

$$\begin{aligned} S^-(k) &= \hat{G}\hat{S}(k) - L^-(k) \\ k &= 1,2,\ldots,M \end{aligned} \tag{1.10B}$$

where $\hat{G}\hat{S}(k)$ is the template and $L^+(k)$ and $L^-(k)$ are functions derived from the variations of the template. These can be taken, for example, to be the estimated variance at each point. Detection is performed as follows: At the time k, an observation vector x(k) is formed from the observation signal $\underline{x}(k)$, such that

$$\underline{x}^T(k) = [x(k), x(k-1), \ldots, x(k-M+1)] \tag{1.11}$$

In the same manner, the upper and lower limit vectors $\underline{S}^+$ and $\underline{S}^-$ are defined:

$$\underline{S}^+ = [S^+(M), S^+(M-1), \ldots, S^+(1)]^T \tag{1.12}$$

$$\underline{S}^- = [S^-(M), S^-(M-1), \ldots, S^-(1)]^T \tag{1.13}$$

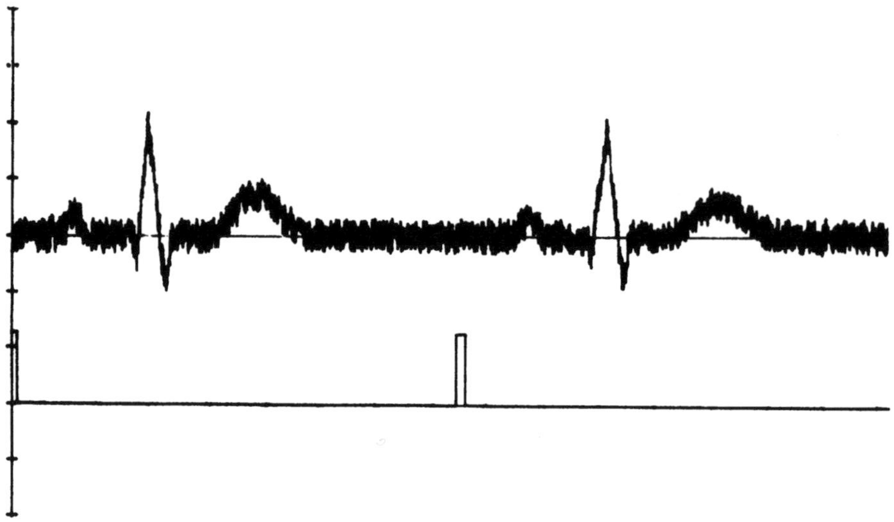

FIGURE 2. Detection of QRS complex by the contour limiting method: upper trace, noisy ECG signal; lower trace, PQRST complex detection.

At each time k, the observation vector is compared with the limits. A wavelet is assumed to be present in the observation window if

$$\underline{S}^- \leq \underline{x}(k) \leq \underline{S}^+ \tag{1.14}$$

If Equation 1.14 is not true, the observation window is assumed not to contain a wavelet. The data in the observation vector is shifted by one sample and the new vector $\underline{x}(k + 1)$ is checked by the Equation 1.14.

No general rules can be given concerning the exact construction of the limit functions $L^+(k)$ and $L^-(k)$. Making these limits large will sometimes allow noise to be recognized as a wavelet (false positive). Making the limits small will cause some wavelets to be rejected (false negative). The decision as to the required safety margin depends on the relative importance of the above two errors to the particular application. Equation 1.14 can be relaxed by requiring that only a certain fraction, say 90%, of the M elements of the observation vector $\underline{x}(k)$ obey Equation 1.14. This will reduce the sensitivity to noise.

Example 1.2 — QRS detection by contour limiting

Consider the signal given by Example 1.1; we shall define the upper and lower limits for the wavelet of Equation 1.6 by

$$\underline{S}^+ = \underline{\tilde{S}}(k) + (0.1\tilde{S}(k) + 0.3)$$

$$\underline{S}^- = \underline{\tilde{S}}(k) - (0.1\tilde{S}(k) + 0.3)$$

and detect the QRS complex of the signal of Example 1.1, by means of Equation 1.14. Figure 2 shows the detection results.

III. MATCHED FILTERING

The problem of detecting wavelets in noise is an important problem in communication theory. Matched filters have been used for optimal detection of Morse codes and digital communications.[33] Let us transfer the observation signal x(t) through a linear filter. We require that the filter will cause the wavelet (when present) to be amplified while relatively attenuating the noise, thus increasing the signal-to-noise ratio. The output of this filter will be subjected to a threshold to detect the presence of a wavelet. Assume that we want to consider an MA filter (Chapter 7, Volume I). The output of the filter is given by the observation vector $\underline{y}^T(k) = [y(k - 1), \ldots, y(k - M + 1)]$, where

$$y(m) = \sum_{j=0}^{M-1} b_j x(m - j) \tag{1.15}$$

We have chosen a filter of order M so that if a wavelet is present, the output of the filter will contain information about the complete wavelet. Consider now the following signal to noise ratio, SNR_o, at the output of the filter.

$$SNR_o = \frac{(E\{y(m)|\underline{x}(m) = \underline{s} + \underline{n}\} - E\{y(m)|\underline{x}(m) = \underline{n}\})^2}{Var\{y(m)|\underline{x} = \underline{n}\}} \tag{1.16}$$

The numerator of Equation 1.16 is the power of the difference between the filter's output with and without a wavelet at the input. The variance of output's noise serves as a normalization factor. Let us arbitrarily choose the *ith* wavelet to represent the time frame including a wavelet, then $m = k_i + M - 1$ and

$$\begin{aligned} E\{y(m)|\underline{x}(m) = \underline{s} + \underline{n}\} &= E\left\{\sum_{j=0}^{M-1} b_j(G_i s_i(M - 1 - j) + n(k_i + M - 1 - j))\right\} \\ &= \sum_{j=0}^{M-1} b_j E\{G_i s_i(M - 1 - j)\} \end{aligned} \tag{1.17}$$

Where we have used the assumption, the noise has a zero mean:

$$E\{y(m)|\underline{x}(m) = \underline{n}\} = 0 \tag{1.18}$$

If we also assume that the noise samples are independent, we get

$$Var\{y(m)|\underline{x}(m) = \underline{n}\} = \sigma_n^2 \sum_{j=0}^{M-1} b_j^2 \tag{1.19}$$

where σ_n^2 is the variance of the noise.

Introducing the last equations into Equation 1.16 yields

$$SNR_o = \frac{\left(\sum_{j=0}^{M-1} b_j E\{G_i s_i(M - 1 - j)\}^2\right)}{\sigma_n^2 \sum_{j=0}^{M-1} b_j^2} \leqslant \frac{\left(\sum_{j=0}^{M-1} b_j^2\right)\left(\sum_{j=0}^{M-1} E\{G_i s_i(M - 1 - j)\}^2\right)}{\sigma_n^2 \sum_{j=0}^{M-1} b_j^2} \tag{1.20}$$

The right term in Equation 1.20 is due to the Schwarz inequality. Maximization of the signal-to-noise ratio of Equation 1.20 is given when equality occurs. This takes place for

$$b_j = KE\{G_i s_i(M - 1 - j)\} \tag{1.21}$$

Let us denote the template $\hat{G}\hat{S}(j)$ by

$$\hat{G}\hat{S}(M - 1 - j) = E\{G_i s_i(M - 1 - j)\} \qquad j = 0,1,\ldots,M - 1 \tag{1.22}$$

Hence the optimal MA filter is given by

$$b_j = K\hat{G}\hat{S}(M - 1 - j) \qquad j = 0,1,\ldots,M - 1 \tag{1.23}$$

Where K is an arbitrary constant, we shall thus choose K = 1. This optimal filter is known as the matched filter.

We can rewrite the filter's coefficients in vector form:

$$\underline{b}^T = [b_0, b_1, \ldots, b_{M-1}] \tag{1.24}$$

and the template's vector:

$$\underline{\hat{S}}^T = [\hat{S}(M - 1), \hat{S}(M - 2), \ldots, \hat{S}(0)] \tag{1.25}$$

Equation 1.23 is then

$$\underline{b} = \hat{G}\underline{\hat{S}} \tag{1.26}$$

and the filter's output from Equation 1.15 is

$$y(m) = \hat{G}\underline{\hat{S}}^T\underline{x}(m) \tag{1.27}$$

where $\underline{x}(m)$ is defined in Equation 1.11. The last equation states that the matched filter is equivalent to a cross correlator, cross correlating the observation window x(m) with the template $\underline{\hat{S}}$. The maximum signal-to-noise ratio for the matched filter is achieved by introducing Equation 1.26 into 1.20.

$$\mathrm{Max}(SNR_o) = \frac{\sum_{j=0}^{M-1} (\hat{G}\hat{S}(M - 1 - j))^2}{\sigma_n^2} = \frac{\hat{G}^2}{\sigma_n^2} \sum_{k=0}^{M-1} \hat{S}^2_{(j)} = \frac{\hat{G}^2}{\sigma_n^2} \underline{\hat{S}}^T\underline{\hat{S}} \tag{1.28}$$

The matched filter procedure can be summarized as follows. We estimate the template $\hat{G}\hat{S}$ and store it in memory. For each new sample of the incoming signal, x(k), we form the observation vector $\underline{x}(k)$ (by opening a window of M samples). We cross correlate the template and observation window to get the k*th* sample of the output. This we compare with the threshold. The observation window for which y(k) has crossed the threshold is considered to contain a wavelet. Correlation-based detection procedures have been applied to biomedical signals.[34-37] Note that here as in the previous discussion, we only determine the presence or absence of a wavelet in the observation window, but not its exact shape.

IV. ADAPTIVE WAVELET DETECTION

A. Introduction

We shall consider now the problem of wavelet detection while estimating and adapting the template.[34] This is required when the *a priori* information is insufficient or when the wavelets are nonstationary and the template has to track the slow variations in wavelets. We consider here a modification of the filter discussed in the previous section.

Consider the average squared error $\epsilon^2(k,M)$ between the signal x(k) and the estimated wavelet:

$$\epsilon^2(k,M) = \frac{1}{M}(\underline{x}(k) - \hat{G}(k)\underline{\hat{S}})^T(\underline{x}(k) - \hat{G}(k)\underline{\hat{S}}) \tag{1.29}$$

The best least squares estimate of the gain, $\hat{G}_0(k)$, is the one that minimizes the error of Equation 1.29. This optimal estimated gain is given by

$$\hat{G}_0(k) = \frac{\frac{1}{M}(\underline{x}^T(k)\underline{\hat{S}})}{\frac{1}{M}(\underline{\hat{S}}^T\underline{\hat{S}})} = \frac{R_{x\hat{s}}}{R_{\hat{s}}} \tag{1.30}$$

where $R_{xy} \triangleq M^{-1}(\underline{x}^T\underline{y})$ is the M window cross correlation.

Note that for a constant template $\underline{\hat{S}}$, the denominator of Equation 1.30, namely, the energy of the template ($R_{\hat{s}}$), is a constant. Introducing Equation 1.30 into Equation 1.29 yields

$$\epsilon^2(k,M) = R_x(k) - \hat{G}_0^2(k)R_{\hat{s}} \tag{1.31A}$$

at time k_j; when the observation vector contains the *jth* wavelet, S_j, Equation 1.30 becomes

$$\epsilon^2(k_j,M) = G_j^2(R_{sj}) - \hat{G}_0^2(k_j)R_{\hat{s}} + 2G_jR_{sn} + R_n \tag{1.31B}$$

For a stationary noise process and large M, the last term of Equation 1.31B is almost constant. Since the noise and wavelets are independent, the third term becomes small. The first two terms of Equation 1.31B denote the error between the energy of the wavelet G_jS_j and its estimate $\hat{G}(k_j)\hat{S}$. This term will yield a minimum at times k_j. Detection of time of occurrence k_j is thus achieved by finding the local minima of the error ϵ^2.

The minima at times k_j are local minima. For example, in the noise-free case, the estimated gain and the error will be zero in segments with no wavelet, while the minimum error at times k_j will be some local minimum, probably above zero. To produce an improved error function we shall introduce a weighting function. This function will ensure higher error values for data associated with low probable gains. Suppose the gains G_j are random variables with known probability distribution P(G). Define a weighting function W(G):

$$W(G) = f(G)\frac{P(E\{G\})}{P(G)} \tag{1.32}$$

The weighted error $\epsilon_w^2(k,M)$ is

$$\epsilon_w^2(k,M) = \epsilon^2(k,M) \cdot W(\hat{G}) \tag{1.33}$$

and is inversely proportional to the gain probability distribution. The function f(G) in Equation 1.32 is chosen such that

$$\lim_{G \to 0} \epsilon_w^2(G) = \infty \tag{1.34}$$

The weighting function assures high errors for very low gains and reduces the error for the more probable ones. Since for low gains the error approaches zero as $\hat{G}^2$, the function f(G) must obey:

$$\lim_{G \to 0} G^2 f(G) = \infty \tag{1.35}$$

For the examples described in this section, the gains were assumed to be gaussian distributed and the function f(G) was chosen as

$$f(G) = \left(\frac{\exp(G/E\{G\} - 1)}{G/E\{G\}}\right)^{\gamma}$$

$$\gamma > 2 \tag{1.36}$$

The parameter γ is heuristically determined; other parameters of Equation 1.36 are determined from *a priori* knowledge of wavelets statistics or by sample estimation during training state. Assuming the correlation window is large enough that $R_{sn}(k) \simeq 0$, it can be shown that the expectation of the weighted error can be approximated by

$$E\{\epsilon_w^2(k,M)\} \simeq \left(G_k \frac{R_{s\hat{s}}}{R_{\hat{s}}}\right)\left(G_k\left(R_s - \frac{(R_{s\hat{s}})^2}{R_{\hat{s}}}\right)\right)$$

$$@\ k = k_j \tag{1.37}$$

The detection of the presence of a wavelet in the signal is performed by placing a threshold on the weighted error function. The value of the threshold level LIM is experimentally determined using a training signal. Assume an initial sample of the analyzed signal as a training signal. This record is analyzed, for example, visually, by a trained person, with L wavelets at times k_i, i = 1,2, . . . ,L, detected. The unweighted error is calculated and its mean is estimated by

$$\hat{E}\{\epsilon^2(k,M)\}_{min} = \frac{1}{L}\sum_{i=1}^{L} \epsilon^2(k_i,M) \tag{1.38}$$

If the range of gains of interest is $G_1 \leq G \leq G_2$ then the threshold level is heuristically determined as

$$LIM = \max(W(G_1),W(G_2)) \cdot \hat{E}\{\epsilon^2(k,M)\}_{min} \tag{1.39}$$

B. Template Adaptation

The need to adapt the template arises in two cases; when the initial information concerning the wavelet is insufficient, and thus $\hat{S}_0$ has to be improved, and when the wavelets are slowly changing in time so that tracking action is required. Template adaption is performed each time a wavelet is detected. Assume that at time m_k the *kth* wavelet was detected; then adaptation is achieved according to

$$\underline{\hat{S}}_k = \rho(k)\underline{\hat{S}}_{k-1} + \psi(k)\underline{x}(m_k) \tag{1.40}$$

where $\underline{\hat{S}}_k$ is the adapted template, $\underline{\hat{S}}_{k1}$ is the previous template, and $\rho(k)$ and $\psi(k)$ are weights. The current template is thus a linear combination of the last template and current observation signal. The k*th* template can be expressed in terms of the (k − N)*th* template:

$$\underline{\hat{S}}_k = \rho_{N-1}(k)\underline{\hat{S}}_{k-N} + \underline{S}_k^w + \underline{n}_k^w$$

$$k \geq N \tag{1.41A}$$

where

$$\rho_x(k) \triangleq \begin{cases} \prod_{j=0}^{x} \rho(k-j) & ; \quad x \geq 0 \\ 0 & ; \quad x < 0 \end{cases} \tag{1.41B}$$

$$\underline{S}_k^w = \sum_{j=0}^{N-1} \rho_{j-1}(k)\psi(k-j)G_{k-j}\underline{S}_{k-j} \tag{1.41C}$$

$$\underline{n}_k^w = \sum_{j=0}^{N-1} \rho_{j-1}(k)\psi(k-j)\underline{n}(m_{k-j}) \tag{1.41D}$$

Equations 1.41A to D express the k*th* template as a combination of a weighted average of the noise vectors associated with the last N wavelets.

The adapted template is a random vector. It is of interest to examine its signal-to-noise ratio and compare it to that of the observation vector. Denote the template's noise power by E_n; then for a stationary uncorrelated noise process having the variance σ_n^2:

$$E_n = \frac{1}{M} E\{(\underline{n}_k^w)^T\underline{n}_k^w\} = \sigma_n^2 \sum_{m=0}^{N-1} \rho_{m-1}^2(k)\psi^2(k-m) \tag{1.42}$$

Denote the template's power by $E_{\hat{s}}$; then

$$E_{\hat{s}} = \frac{1}{M} E\{(\underline{S}_k^w + \rho_{N-1}(k)\underline{\hat{S}}_{k-N})^T(\underline{S}_k^w + \rho_{N-1}(k)\underline{\hat{S}}_{k-N})\} \tag{1.43}$$

The template's signal-to-noise ratio is given by:

$$SNR_{\hat{s}} = \frac{E_{\hat{s}}}{\sigma_n^2 \sum_{m=0}^{N-1} \rho_{m-1}^2(k)\psi^2(k-m)} \tag{1.44}$$

The signal-to-noise ratio of the observation vector at time m_k is similarly given by:

$$SNR_x = \frac{E\{G_k^2\}\underline{S}_k^T\underline{S}_k}{E\{\underline{n}^T\underline{n}\}} = \frac{R_s(m_k)E\{G_k^2\}}{\sigma_n^2} \tag{1.45}$$

The ratio $SNR_{\hat{s}}/SNR_x$ given by Equations 1.44 and 1.45 is a measure of the relative noisiness of the adapted template. We shall now deal separately with the two cases of adaptation.

C. Tracking a Slowly Changing Wavelet

In this case, the adaptation must be performed continuously; since no information on the time course of change is given, constant weights will be used. Denote $\rho(k) = \rho$; $\psi(k) = \psi$ and assume slow variations such that for the last N wavelets, the following assumptions can be made

$$G_j \simeq G; \quad \underline{\tilde{S}}_j \simeq \underline{S} \text{ for all } j \in (k - N, k) \tag{1.46}$$

Since only slow changes in wavelets are allowed, one can assume "almost" stationary conditions with

$$E\{\underline{\hat{S}}_{k-1}\} \simeq E\{\underline{\hat{S}}_k\} \simeq E\{\underline{S}_k\} \tag{1.47}$$

Taking the expectation of Equation 1.40, with $E\{n\} = 0$ and the assumption of Equation 1.47 yields

$$\psi = (1 - \rho)/G \tag{1.48}$$

The k*th* estimate of the template is given from Equations 1.41 and 1.46:

$$\underline{\hat{S}}_k = \rho^N_{\underline{\hat{S}}_{k-N}} + \psi G\underline{S}\frac{1 - \rho^{N+1}}{1 - \rho} + \psi\sum_{j=0}^{N-1}\rho^j\underline{n}_{k-j}$$

$$0 \leq \rho < 1 \tag{1.49}$$

The relative noisiness of the estimate is given from Equations 1.44 to 1.46 by

$$\frac{SNR_{\hat{s}}}{SNR_x} = \frac{(\rho^N - 1)(\rho + 1)}{(\rho^N + 1)(\rho - 1)} + \frac{2}{\psi G}\frac{R_{s\hat{s}_{K-N}}}{R_s}\frac{\rho^N(\rho + 1)}{\rho^N + 1} + \frac{1}{\psi^2 G^2}\frac{R_{\hat{s}_{K-N}}}{R_s}\frac{\rho^{2N}(\rho^2 - 1)}{\rho^{2N} - 1}$$

$$0 \leq \rho < 1 \tag{1.50}$$

In selecting the adaptation coefficients ρ and ψ, both the tracking rate and noisiness of the template have to be considered. Define the tracking coefficient T_c to be the ratio between the template ($G\underline{S}$) part and the initial estimate ($\underline{\hat{S}}_{k-N}$) part of Equation 10.49, hence

$$T_c = \psi G\frac{1 - \rho^{N+1}}{1 - \rho}\rho^{-N}$$

$$0 \leq \rho < 1 \tag{1.51}$$

In order to optimally select the adaptation coefficients, we shall define a cost function, I_c:

$$I_c = \ln(T_c)\frac{SNR_{\hat{s}} - SNR_x}{SNR_x} \tag{1.52}$$

The logarithm of T_c is taken in order to equalize order of magnitude of parameters. Maximization of Equation 1.52, together with Equation 1.40, provides the optimal adaptation coefficients.

D. Correction of Initial Template

In this case, the wavelets are assumed to be constant. It is required to improve the initial estimate of the template. Here, the adaptation coefficients are time dependent. Exponential decay has been chosen such that

$$\rho(k) = 1 - (1 - \rho_0)e^{-\alpha k} \quad ; \quad 0 \leq \rho_0 \leq 1$$

$$\psi(k) = \psi_0 e^{-\alpha k} \quad ; \quad 0 \leq \psi_0 \leq 1 \qquad (1.53)$$

The adapted template is thus a weighted average of about $K = 5/\alpha$ initial templates. After this period, $\psi(k) \simeq 0$ and the adaptation process is terminated. *A priori* knowledge and estimation of the goodness of the initial template $\underline{\hat{S}}_0$ allows the determination of α.

In this case, the convergence rate coefficient C_c will be defined instead of the tracking coefficient (Equation 1.51).

$$C_c(k) = \frac{\Delta E(0) - \Delta E(k)}{R_s} \qquad (1.54)$$

where

$$\Delta E(k) = \frac{1}{M}\{(\underline{\hat{S}}_k - \underline{S})^T(\underline{\hat{S}}_k - \underline{S})\} \qquad (1.55)$$

$\Delta E(k)$ is thus the mean square error between the current template and the wavelet. The relation between the adaptation coefficients α,ρ_0,ψ_0 is given by the maximization of the cost function

$$I_c = C_c \frac{SNR_{\hat{s}}(k) - SNR_x}{SNR_x} \qquad (1.56)$$

Note that here, the assumption in Equation 1.47 can be used only with great care since templates may be highly nonstationary.

Example 1.3 — adaptive wavelet detection of QRS complexes

Consider the problem of detecting the QRS complex in a noisy ECG. In this example the ECG signal was sampled at 2 kHsz and a template of M = 512 samples was used. The template's duration is thus 256 m sec which is more than sufficient for the QRS complex. The real ECG signal was corrupted with additive white noise. Figure 3 shows the signal, the template, and the weighted error for ECG with various signal-to-noise ratios. The template and the parameters for the Gaussian distribution were estimated from a sample of 13 QRS detected manually. Detection was performed here with no template adaptation.

To check the adaptation algorithm, consider three types of initial templates: a monophasic positive triangle, biphasic positive and negative triangles, and triphasic negative-positive-negative triangle templates. Figure 4 shows the adaptation of the templates and the error for the first five wavelets.

Consider now the case where the noise consists of pulsative interferences and base line shifts. To simulate such interferences, the ECG was contaminated by a train of random pulses, the height and width of which were uniformally distributed in the range of $\pm 10\%$ of the average R wave amplitude and width. Pulse interferences were placed between each two successive QRSs. To demonstrate base line shifts, a slow sine wave was added to the ECG. Figure 5 shows the results.

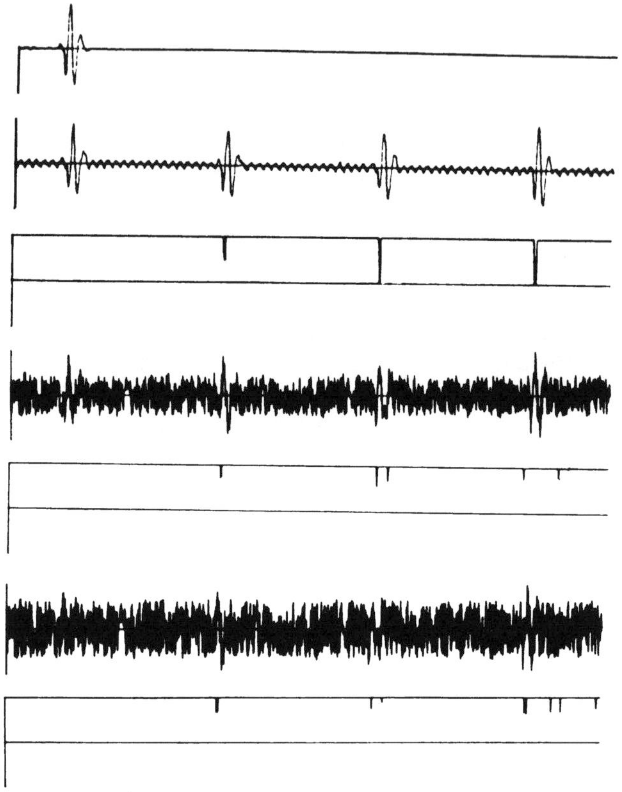

FIGURE 3. Detection error for real ECG signal. (A) Template; (B), (C) signal and weighted error, SNR = infinity; (D), (E) signal and weighted error, SNR = 2.48; (F), (G) signal and weighted error, SNR = 1.24. (From Cohen, A. and Landsberg, D., *IEEE Trans. Biomed. Eng.*, BME-30, 332, 1983 (© 1983, IEEE). With permission.)

V. DETECTION OF OVERLAPPING WAVELETS

A. Statement of the Problem

The assumption made in Equation 1.3 that wavelets do not overlap may be too restrictive for some applications. Consider, for example, the problem of multispike train analysis where an electrode is used to record an action potential from a neuron or a muscle. Very often neighboring neurons fire at the same time and generate several overlapping wavelets. Another example may be the recording of visual evoked responce potential, VER.[11] The VER is assumed to be an aggregate of overlapping wavelets, generated by multiple spatially disparate sources. The various wavelets are unknown in their exact shape and timing. Dye dilution curves (see Appendix A) also contain overlapping wavelets that are interfering with the detection of the main wavelet.

We shall follow De Figueiredo and Gerber[37] in formulating and solving the overlapping problem. We shall remove the restriction Equation (1.3) and replace it by the assumption that in any given observation window the same wavelet s_i can appear no more than once; we allow, however, n different wavelets to overlap in the window. Hence

$$x(t) = \sum_{i=1}^{n} G_i s_i(t - t_i) + n(t)$$

$$t \epsilon (t, t - T) \qquad (1.57)$$

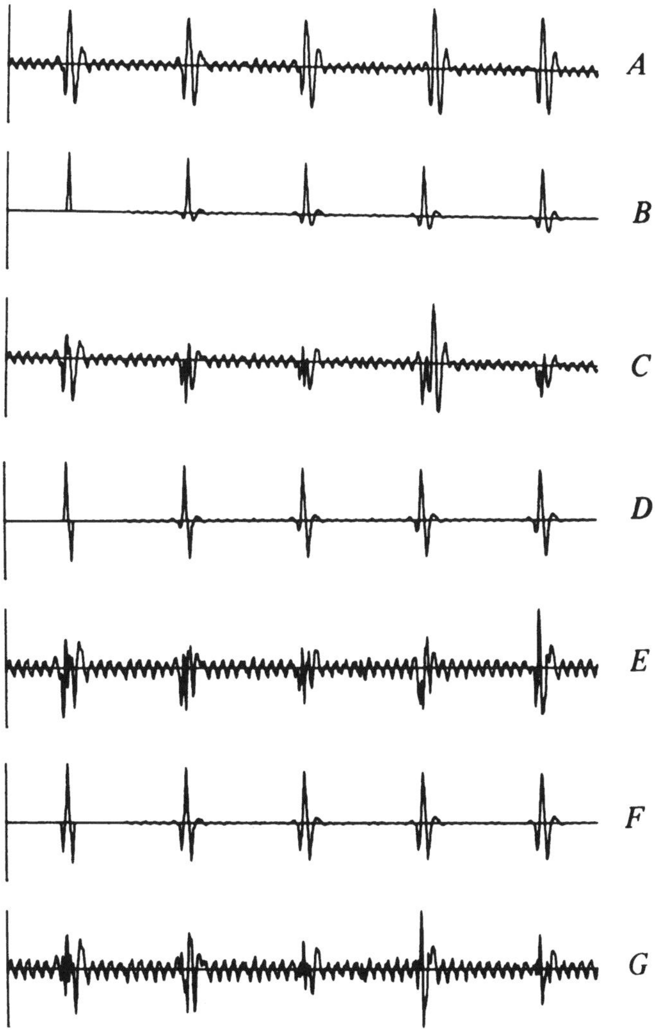

FIGURE 4. Adaptation of templates. (A) Signal; (B), (C) monophasic template and corresponding weighted error; (D), (E) biphasic template; (F), (G) trisphasic template (errors not plotted to scale). $\rho_0 = 0.8$; $\alpha = 0.5$; $\psi_0 = (1 - \rho_0)/G$. (From Cohen, A. and Landsberg, D., *IEEE Trans. Biomed. Eng.*, BME – 30, 332, 1983 (© 1983, IEEE). With permission.)

as before, the G_i's and t_i's are unknown and n(t) is white noise. De Figueiredo's algorithm is implemented in two steps.

B. Initial Detection and Composite Hypothesis Formulation

Consider n matched filters. Rather than using the correlation filters, DeFigueiredo suggested much simpler "area" filters. The output of the *jth* filter, $z_j(\tau)$, is

$$z_j(\tau) = \int_T |\hat{G}_j\hat{s}_j(t - \tau) - x(t)|dt$$
$$j = 1,2,\ldots,n \qquad (1.58)$$

where $\hat{G}_j\hat{S}_j$ is the *jth* template and most probable gain.

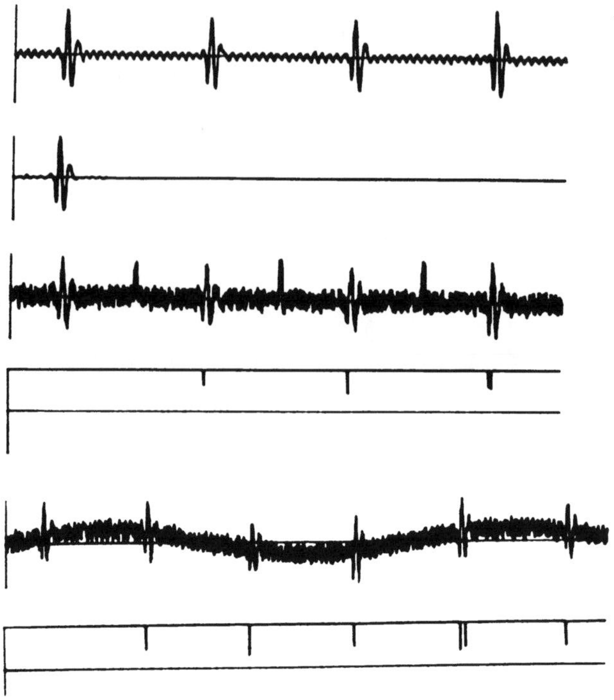

FIGURE 5. Rejections of interferences from real ECG record. (A) Signal; (B) template; (C), (D) signal with pulsative interferences and additive noise and corresponding weighted error; (E), (F) signal with low frequency sine wave interference and corresponding weighted error. (From Cohen, A. and Landsberg, D., *IEEE Trans. Biomed Eng.*, BME-30, 332, 1983 (© 1983, IEEE). With permission.)

The output of the filter is achieved by integrating (over the window) the absolute value of the difference between the observation and the template. This procedure is repeated while shifting the template with various delays, τ. Note that these filters require no multiplications. The point where $z_j(\tau)$ gets its minimum is the most likely location (τ) of S_j in the composite wavelet and serves as the first estimate for τ_j, and the actual participation of S_j. The minima points of all $z_j(\tau)$, $j = 1,2, \ldots ,n$, are compared with a threshold. All templates whose corresponding filters provide minimum below the threshold are hypothesized to be present in the composite wavelet.

C. Error Criterion and Minimization

The gains G_i and delay τ_i are estimated by minimizing the performance index

$$J(G,\tau) = \frac{1}{2} \int \left| x(t) - \sum_{i=1}^{n} \hat{G}_i \hat{s}_i(t - \tau_i) \right|^2 dt \qquad (1.59)$$

$J(G,\tau)$ is differentiated with respect to G_i and τ_i and the result is set equal to zero. This leads to the set of 2n normal equations

$$\sum_{i=1}^{n} R_{s_j s_i}(\tau_j - \tau_i) \cdot G_i = R_{s_j x}(\tau_j)$$

$$j = 1,2,\ldots,n \qquad (1.60A)$$

$$\sum_{i=1}^{n} R_{s_j s_i}(\tau_j - \tau_i) \cdot G_i = R_{s_j x}(\tau_j)$$

$$j = 1,2,\ldots,n \tag{1.60B}$$

where

$$R_{ab}(\tau) = \int a(\lambda)b(\lambda + \tau)d\lambda \tag{1.61}$$

and $\dot{S}_j$ is the time derivative of S_j.

Equations 1.60A and B cannot be solved explicitly; De Figueiredo has solved the minimization by a variable metric algorithm. The interested reader is referred to his paper.[37]

REFERENCES

1. **Godfrey, K. R. and Bruce, D. M.,** The identification of isolated events in electroencephalograms, in *Identification and System Parameter Estimation,* Rajbman, Ed., North-Holland, New York, 1978, 549.
2. **Lim, A. J. and Winters, W. D.,** A practical method for automatic real time EEG sleep state analysis, *IEEE Trans. Biomed. Eng.*, 27, 212, 1980.
3. **Smith, J. R.,** Automatic analysis and detection of EEG spikes, *IEEE Trans. Biomed. Eng.*, 21, 1, 1974.
4. **Frost, J. D.,** Microprocessor based EEG spike detection and quantification, *Int. J. Bio-Med. Comput.*, 10, 357, 1979.
5. **Ktonas, P. Y., Luoh, W. M., Kejariwal, M. L., Reilly, E. L., and Seward, M. A.,** Computer aided quantification of EEG spike and sharp wave characteristics, *Electroencephalogr. Clin. Neurophysiol.*, 51, 237, 1981.
6. **Saltzberg, B., Lustick, S., and Heath, R. G.,** Detection of focal depth spiking in the scalp EEG of monkeys, *Electroencephalogr. Clin. Neurophysiol.*, 31, 327, 1971.
7. **Bremer, G., Smith, J. R., and Karacan, I.,** Automatic detection of the K-complex in sleep electroencephalograms, *IEEE Trans. Biomed. Eng.*, 17, 314, 1970.
8. **Smith, J. R., Negin, N., and Nevis, A. H.,** Automatic analysis of sleep EEG's by the hybrid computation, *IEEE Trans. Syst. Sci. Cybern.*, 5, 278, 1969.
9. **Widrow, B.,** The "Rubber Mask" technique. I. Pattern measurements and analysis, *Pattern Recognition,* 5, 175, 1973.
10. **Giaquinto, S. and Marciano, F.,** Automatic stimulation triggered by EEG spindles, *Electroencephalogr. Clin. Neurophysiol.*, 30, 151, 1970.
11. **Senmoto, S. and Childers, D. G.,** Adaptive decomposition of a composite signal of identical unknown wavelets in noise, *IEEE Trans. Syst. Man. Cybern.*, 2, 59, 1972.
12. **Abeles, M. and Goldstein, M. H., Jr.,** Multispike train analysis, *Proc. IEEE,* 65, 762, 1977.
13. **Sanderson, A. C.,** Adaptive filtering of neuronal spike train data, *IEEE Trans. Biomed. Eng.*, 27, 271, 1980.
14. **D'Hollander, E. H. and Orban, G. A.,** Spike recognition and online classification by unsupervised learning system, *IEEE Trans. Biomed. Eng.*, 26, 279, 1979.
15. **Cashman, P. M. M.,** A pattern recognition program for continuous ECG processing in accelerated time, *Comput. Biomed. Res.*, 11, 255, 1980.
16. **Azevedo, S. and Longini, R. L.,** Abdominal lead fetal Electrocardiographic R-wave enhancement for heart rate determination, *IEEE Trans. Biomed. Eng.*, 27, 255, 1980.
17. **Brodda, K., Wellner, U., and Mutschler, W.,** A new method for detection of P waves in ECG's, *Signal Process.*, 1, 15, 1979.
18. **Iwata, A. N., Ishii, N., and Suzumnra, N.,** Algorithm for detecting the first and the second heart sounds by spectral tracking, *Med. Biol. Eng. Comput.*, 18, 19, 1980.
19. **Cohen, A. and Landsberg, D.,** Analysis and automatic classification of breath sounds, *IEEE Trans. Biomed. Eng.*, 31, 585, 1984.
20. **Pavlidis, T.,** Algorithms for shape analysis of contours and waveforms, *IEEE Trans. Pattern Anal. Mach. Intelligence,* 2, 301, 1980.

21. **Lewis, J. W. and Graham, A. H.,** High speed algorithms and damped spline regression and electrocardiogram feature extraction, paper presented at the IEEE Workshop on Pattern Recognition and Artificial Intelligence, Princeton, N.J., 1978.
22. **Holsinger, W. P., Kempner, K. M., and Miller, M. H.,** A QRS processor based on digital differentiation, *IEEE Trans. Biomed. Eng.*, 18, 212, 1971.
23. **De Vries, J., Wisman, T., and Binnie, C. D.,** Evaluation of a simple spike wave recognition system, *Electroencephalogr. Clin. Neurophysiol.*, 51, 328, 1981.
24. **Goldberger, A. L. and Bhargava, V.,** Computerized measurement of the first derivative of the QRS complex: theoretical and practical considerations, *Comput. Biomed. Res.*, 14, 464, 1981.
25. **Haywood, L. J., Murthy, V. K., Harvey, G., and Saltzberg, S.,** On line real time computer algorithm for monitoring ECG waveforms, *Comput. Biomed. Res.*, 3, 15, 1970.
26. **Fischhof, T. J.,** Electrocardiographic diagnosis using digital differentiation, *Int. J. Bio-Med. Comput.*, 13, 441, 1982.
27. **Colman, J. D. and Bolton, M. P.,** Microprocessor detection of electrocardiogram R-waves, *J. Med. Eng. Technol.*, 3, 235, 1979.
28. **Talmon, J. L. and Hasman, A.,** A new approach to QRS detection and typification, *IEEE Comput. Cardiol.*, 479, 1981.
29. **Nygards, M. E. and Sornmo, L.,** Delineation of the QRS complex using the envelope of the ECG, *Med. Biol. Eng. Comput.*, 21, 538, 1983.
30. **Uijen, G. J. H., De Weerd, J. P. C., and Vendrik, A. J. H.,** Accuracy of QRS detection in relation to the analysis of high frequency components in the ECG, *Med. Biol. Eng. Comput.*, 17, 492, 1979.
31. **Van den Akker, T. J., Ros, H. H., Koelman, A. S. M., and Dekker, C.,** An on-line method for reliable detection of waveforms and subsequent estimation of events in physiological signals, *Comput. Biomed. Res.*, 15, 405, 1982.
32. **Goovaerts, H. G., Ros. H. H., Van den Akker, T. J., and Schneider, H. A.,** A digital QRS detector based on the principle of contour limiting, *IEEE Trans. Biomed. Eng.*, 23, 154, 1976.
33. **Papoulis, A.,** *Signal Analysis*, McGraw-Hill, Kogakusha, Auckland, 1981.
34. **Cohen, A. and Landsberg, D.,** Adaptive real time wavelet detection, *IEEE Trans. Biomed. Eng.*, 30, 332, 1983.
35. **Collins, S. M. and Arzbaecher, R. C.,** An efficient algorithm for waveform analysis using the correlation coefficient, *Comput. Biomed. Res.*, 14, 381, 1981.
36. **Fraden, J. and Neuman, M. R.,** QRS wave detection, *Med. Biol. Eng. Comput.*, 18, 125, 1980.
37. **De Figueiredo, R. J. P. and Gerber, A.,** Separation of superimposed signals by a cross-correlation method, *IEEE Trans. Acoust. Speech Signal Process.*, 31, 1084, 1983.
38. **Cox, J. R., Nolle, F. M., Fozzard, H. A., and Oliver, G. C.,** AZTEC: a preprocessing program for real time ECG rhythm analysis, *IEEE Trans. Biomed. Eng.*, 15, 128, 1968.
39. **Abenstein, J. P. and Tompkins, W. J.,** A new data reduction algorithm for real-time ECG analysis, *IEEE Trans. Biomed. Eng.*, 29, 43, 1982.

Chapter 2

POINT PROCESSES

I. INTRODUCTION

Point processes[1-3] are random processes which produce random collections of point occurrences, or series of events, usually (but not necessarily) along the time axis. In univariate point process analysis, the exact shape of the event is of no interest. The "time" of occurrence (or the intervals between occurrences) is the only information required. A more general case is the multivariate point process in which several classes of points are distinguished. In the multivariate case, the shape of the event serves only to classify it. The statistics of the process, however, are given in terms of the intervals only. Point processes can be viewed as a special case of general random processes[4] and can be dealt with as a type of time series. Point processes analysis has been applied to a variety of applications ranging from the analysis of radioactive emission to road traffic studies and to queuing and inventory control problems. Point processes theory has been applied to the analysis of various biomedical signals.[4-6] The main application, however, has been in the field of neurophysiology.[7-16]

A neural spike train is the sequence of action potentials picked up by an electrode from several neighboring neurons. The neurophysiologist is interested in the underlying cellular mechanisms producing the spikes. He may investigate, for example, the effects of environmental conditions such as temperature, pressure, or various ion concentrations or the effects of pharmacological agents. The analysis of the spike train may be used for the description, comparison, and classification of neural cells. Different interval patterns may result from the same cell under different conditions. Interneural connections may be investigated by analyzing the corresponding spike trains. Multivariate point processes analysis is sometimes required[10] when the spike train contains action potentials from more than one neuron. Classification of each spike into one of the classes to be considered in the analysis is needed. Classification of spikes is done by means of the methods discussed in Chapter 1, Volume II. Figure 1 shows a record of neural spike train.

Analysis of myoelectric activities has also been performed by point processes methods.[17-19] Here the motor unit action potential train has been modeled as a point process. Characteristic deviations from normal motor unit firing patterns were suggested to serve as a diagnostic tool in neuromuscular diseases.[20] It was found, for example, that both firing rate and SD of the interpotential intervals increase in patients with myopathy.

The ECG signal can be considered a point process[21,22] when only the rhythm of the heartbeat is of interest and not the detailed time course of polarization and depolarization of the heart muscle. The occurrence of the R wave is defined as an event and the R-R interval statistics is of interest. Figure 2 shows a record of ECG signal. The high signal-to-noise ratio allows the detection of R waves with a simple threshold device thus generating a point process record.

The occurrence of glottal pulses during voiced segments of speech[23] can be analyzed as a point process. The time interval between consecutive glottal pulses, known as the pitch period, is a function of the vocal cord's anatomy. Laryngial disorders can be diagnosed[24] by means of speech signal analysis. Here the detection of the event is not an easy task. Several algorithms have been suggested[23] for pitch extraction. Figure 3 shows a sample of voiced speech where the pitch events are clearly seen.

Once the events of the process have been defined, data is fitted[25] into a point process model. The most often used models are the renewal process, Poisson distribution, Erlang (Gamma) distribution, Weiball distribution, and AR and MA processes. Analysis of the

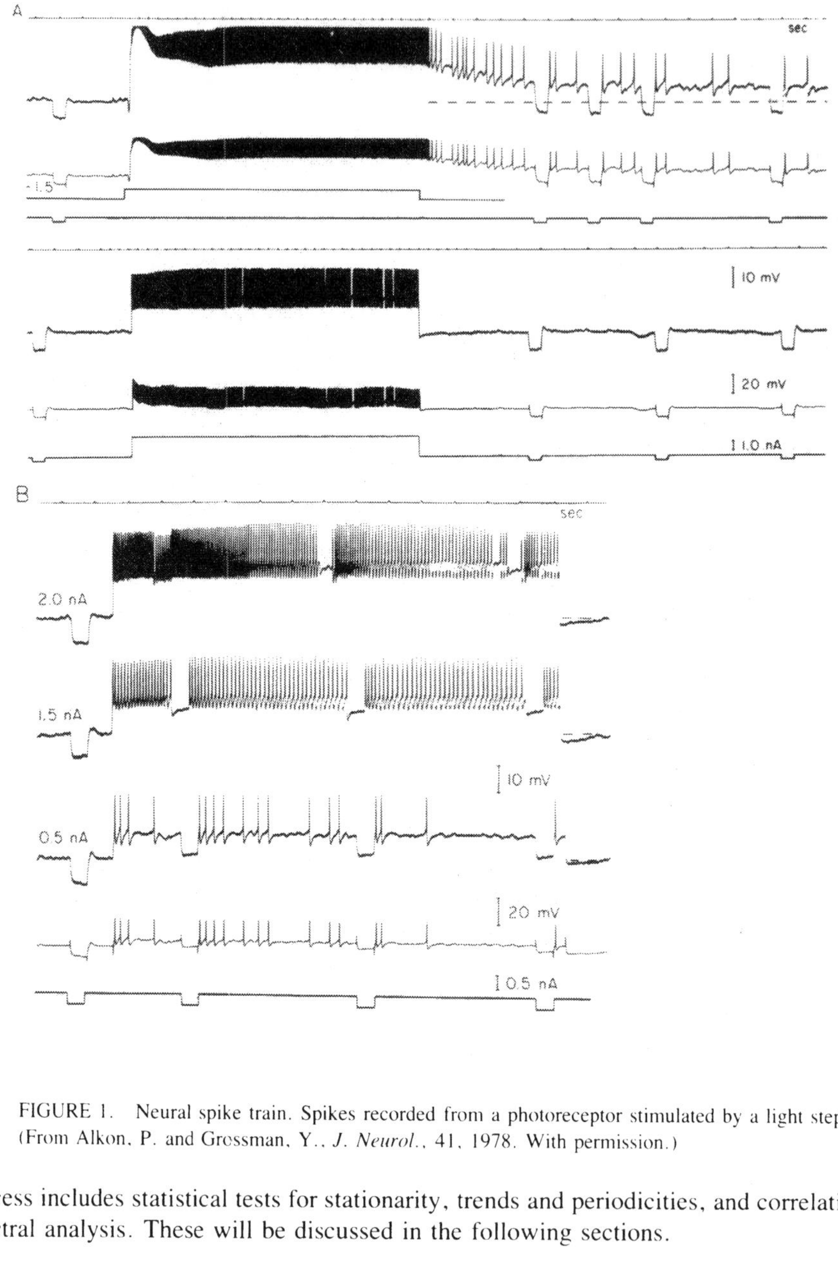

FIGURE 1. Neural spike train. Spikes recorded from a photoreceptor stimulated by a light step. (From Alkon, P. and Grossman, Y., *J. Neurol.*, 41, 1978. With permission.)

process includes statistical tests for stationarity, trends and periodicities, and correlation and spectral analysis. These will be discussed in the following sections.

II. STATISTICAL PRELIMINARIES

The point process is completely characterized by one or two of the canonical forms:[1] the interval process and the counting process. These are schematically described in Figure 4.

The interval process describes the time behavior of the events. The random times, t_i, i = 1,2, . . . ,M, at which the *ith* event occurs is a way to describe the process. Here an arbitrary point in time is chosen as a reference. At this origin point, an event may or may not have occurred. The time intervals between two adjacent events, T_i, i = 1,2, . . . ,M − 1, can also describe the process.

Of interest also are the higher-order intervals. The *nth* order interval is defined as the

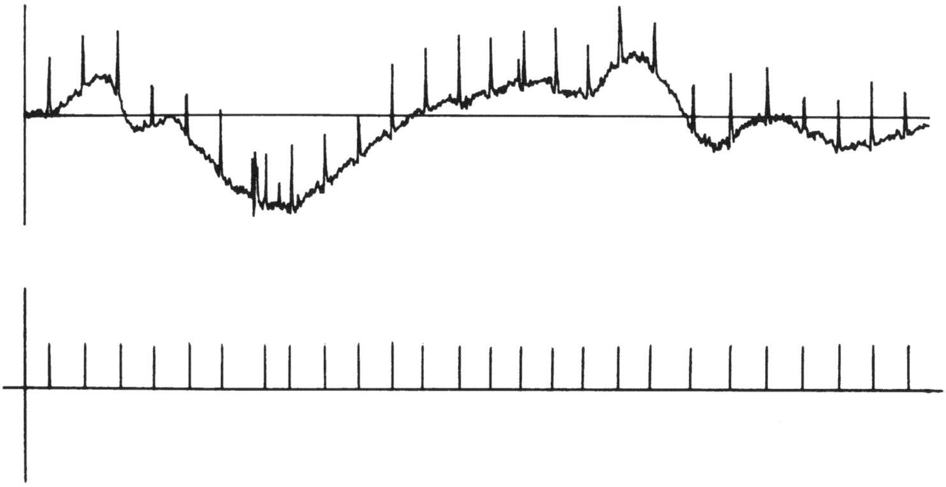

FIGURE 2. ECG signal presented as a point process.

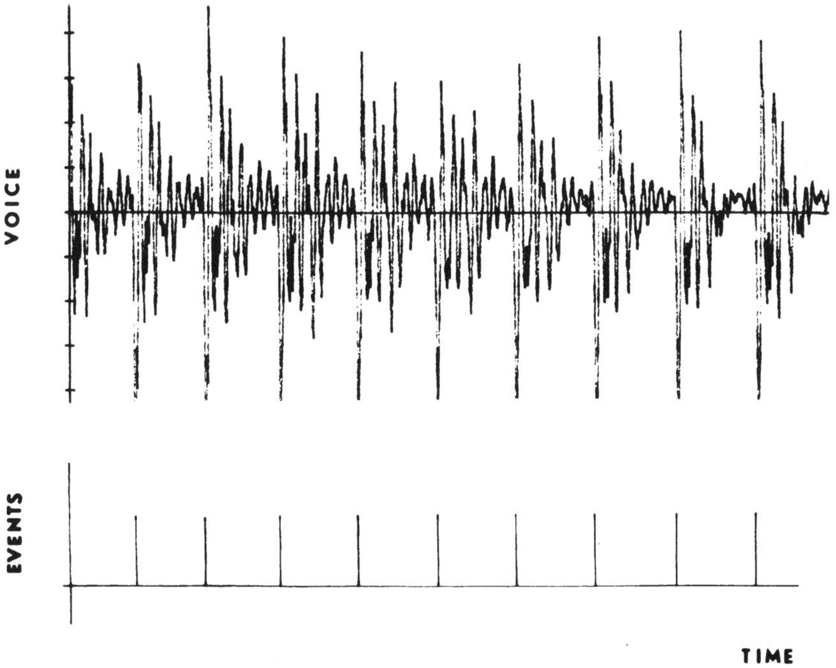

FIGURE 3. Voiced speech signal presented as a point process.

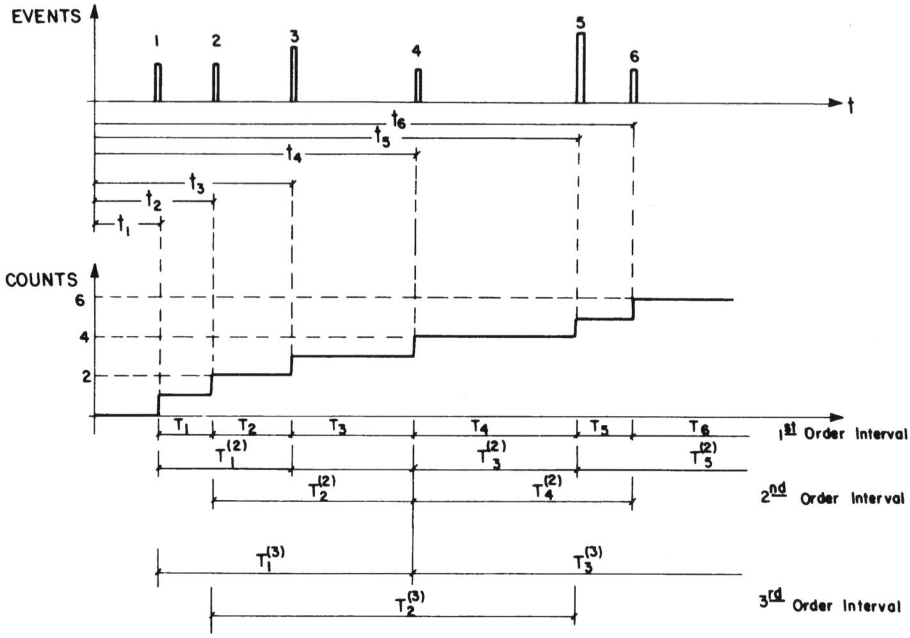

FIGURE 4. Events train.

elapsed time between an event and the *nth* following event. Denote the *nth* order interval by $T_i^{(n)}$, then:

$$T_i^{(n)} = \sum_{j=0}^{n-1} T_{i+j}$$

$$i = 1,2,\ldots \qquad (2.1)$$

We shall define the quantity $N(t_1,t_2)$ to be the random number of events in the range (t_1,t_2). We shall require that there are essentially no multiple simultaneous occurrences, namely, the following condition exists:

$$\lim_{\Delta t \to 0} (\text{Prob}\{N(t,t + \Delta t) > 1\}) = 0(\Delta t) \qquad (2.2)$$

A process for which Equation 2.2 holds is called an "orderly process". The random quantity N(o,t) yields the counting canonical form of the process. The various random variables defined above are drawn from the underlying probability distribution of the point process under test. Their statistics, usually first- and second-order statistics, are used to characterize the investigated process.

Note that for the case of higher-order intervals, as the order n becomes larger there is substantial overlapping of the original intervals. The central limit theorem suggests that for most distributions of the original interval, the *nth* order interval distribution will tend toward gaussian.

The random variable t_i (or T_i) is described by one of several equivalent functions. The probability density function, p(t), describing the random variable t_i, is defined such that $p(t)\cdot\Delta t$ is the probability that an event occurs between t and $t + \Delta t$. The probability density function (PDF) may be expressed as:

$$p_i(T) = \lim_{\Delta T \to 0} \left(\frac{\text{Prob}\{T < T_i \leq T + \Delta T\}}{\Delta T} \right) \tag{2.3}$$

with:

$$\int_0^{\infty} p_i(T)dT = 1 \tag{2.4}$$

The interval histogram is often used as an estimator for the interval PDF. The cumulative distribution function, $P_i(T)$, is the probability that the random variable T_i is not greater than T, hence,

$$P_i(T) = \text{Prob}\{T_i \leq T\} = \int_0^{T} p_i(\tau)d\tau \tag{2.5}$$

The probability that the random variable is indeed greater than T is termed the survivor function, $R_i(T)$:

$$R_i(T) = \text{Prob}\{T_i > T\} = 1 - P_i(T) = \int_T^{\infty} p_i(T)dT \tag{2.6}$$

Sometimes the logarithmic survivor function is used refering to the $\ln(R_i(T))$.

Another function that is sometimes used is the hazard function, $\phi(t,\tau)$. The hazard function is defined such that $\phi(t,\tau)\Delta t$ is the probability that an event has occurred in the interval $(t,t + \Delta t)$ given that the previous event has occurred at time τ. Hence,

$$\phi(t,\tau) = \lim_{\Delta t \to 0} \left(\frac{\text{Prob}\{N(t,t + \Delta t) > 0|(\text{last event occurred at } \tau)\}}{\Delta t} \right) \qquad \tau < t \tag{2.7}$$

The hazard function is also known as the "postevent probability", "age specific failure rate", "conditional probability", or "conditional density function". The hazard function may be constant (as in the Poisson process) or may vary with τ. Pacemaker neurons, for example, exhibit interspike interval distributions with positive hazard function. Some neurons in the auditory system, for example, exhibit interval distributions with negative hazard functions. A similar function is the "intensity function". The complete intensity function,[3] $h_o(t)$, is defined as:

$$h_o(t) = \lim_{\Delta t \to 0} \left(\frac{\text{Prob}\{N(t,t + \Delta t)\} > 0}{\Delta t} \right) \tag{2.8}$$

The conditional intensity function, $h(\tau)$, is defined such that $h(\tau)\Delta t$ is the probability that an event has occurred at time $(t + \tau)$ given that an event has also occurred at time t, hence,

$$h(\tau) = \lim_{\Delta t \to 0} \left(\frac{\text{Prob}\{N(t + \tau,t + \tau + \Delta t) > 0|N(t,t + \Delta t) > 0\}}{\Delta t} \right) \tag{2.9}$$

Note that for a stationary process the conditional intensity function is not a function of t. Note also that the difference between Equations 2.7 and 2.9 is that the hazard function is conditioned upon having the previous event at τ, namely, no event has occurred in the interval (τ,t), while the intensity function is conditioned only to the occurrence of an event at τ.

The point process can also be described by means of the counting process (Figure 4). The counting process, N(t), represents the cumulative number of events in the time interval (0,t). Hence,

$$N(t) = N(0,t) \tag{2.10}$$

The relationship between the two forms, the counting and interval form, is as follows:[14]

$$N(t) < i \text{ if and only if } t_i = \sum_{k=1}^{i} T_k > t \tag{2.11}$$

Equation 2.11 states that at all times smaller than t_i, the cumulative event counts must be smaller than i. This is true since no simultaneous events are allowed (Equation 2.2).

Equation 2.11 yields (using Equation 2.6):

$$\text{Prob}\{N(t) < i\} = \text{Prob}\{t_i > t\} = R_i(t) = 1 - P_i(t) \tag{2.12}$$

hence,

$$\text{Prob}\{N(t) = i\} = P_i(t) - P_{i+1}(t) \tag{2.13}$$

and also,

$$P_i(t) = 1 - \sum_{q=0}^{i-1} \text{Prob}\{N(t) = q\} = \sum_{q=i}^{\infty} \text{Prob}\{N(t) = q\} \tag{2.14}$$

The last equations show that a direct relationship between the counting and interval forms exists. The two processes are equivalent only by way of their complete probability distributions.[1] In usual practice the analysis is based only on the first- and second-order properties of the process. Such an analysis, based on the first and second order of a counting process, is not equivalent to the analysis based on the interval process and information is gained by considering both forms.

III. SPECTRAL ANALYSIS

A. Introduction

In general, the intervals (counts or event times) are statistically dependent. Hence the joint PDF, $p(T_1,T_2, \ldots ,T_n)$, rather than Equation 2.3 has to be considered. The dependency is usually experimentally analyzed by means of joint interval histograms (or scattering diagrams) where two-dimensional plots describing the relations between $p(T_i)$ and $p(T_{i+j})$ are given.

The second-order statistics are very often analyzed by means of the correlation and power spectral density functions. In the analysis of point processes, two different types of frequency domains have been introduced, that of the intervals and that of the event counts.

B. Interevent Intervals Spectral Analysis

The relationships between interevent intervals can be measured by means of the scattering diagrams discussed before. Another quantitative measure is the measure known as "serial correlation coefficients'' of interval lengths[11,13] which is indeed the normalized autocovariance[3] of the process.

Denote the k*th* covariance of the event interval, C_k; then,

$$C_k = \text{Cov}\{T_i, T_{i+k}\} = E\{(T_i - \mu_T)(T_{i+k} - \mu_T)\}$$
$$k = \ldots, -1, 0, 1, \ldots \tag{2.15}$$

where $\mu_T = E\{T\}$ is the expectation of the stationary interval process. The expectation operator in Equation 2.15 means integration over the joint PDF. Let the variance of the interval process be

$$\sigma_T^2 = E\{(T - \mu_T)^2\} \tag{2.16}$$

The serial correlation coefficient, ρ_k, is the normalized autocovariance given by Equations 2.15 and 2.16:

$$\rho_k = \frac{C_k}{\sigma_T^2}$$
$$k = \ldots, -1, 0, 1, \ldots \tag{2.17}$$

The sequence $\{\rho_k\}$ is known as the serial correlogram. It is easily shown that $-1 \leq \rho_k \leq 1$. The serial correlation coefficients have been used extensively to describe statistical properties of neural spike intervals. In practice the serial correlation coefficients have to be estimated from a finite sample with N intervals. A commonly used estimate[15] for ρ_k is

$$\hat{\rho}_k = \frac{\sum_{i=1}^{N} (T_i - \hat{\mu}_T(0))(T_{i+k} - \hat{\mu}_T(k))}{\left[\sum_{i=1}^{N} (T_i - \hat{\mu}_T(0))^2 \cdot \sum_{i=1}^{N} (T_{i+k} - \hat{\mu}_T(k))^2\right]^{1/2}} \tag{2.18A}$$

with:

$$\hat{\mu}_T(k) = \frac{1}{N} \sum_{i=1}^{N} T_{i+k} \tag{2.18B}$$

The interval power spectral density (PSD), $S_I(w)$, is given by the Fourier transform of the serial correlation; hence:

$$S_I(w) = \frac{\sigma_T^2}{2\Pi} \sum_{k=-\infty}^{\infty} \rho_k \exp(-jkw) = \frac{\sigma_T^2}{2\Pi}\left(1 + 2\sum_{k=1}^{\infty} \rho_k \cos(kw)\right) \tag{2.19}$$

where σ_T^2 is the interval variance (Equation 2.16).

The estimation of the PSD function is discussed in Chapter 8, Volume I. The PSD function is used as a test of independence and to compare several point processes.

C. Counts Spectral Analysis

Let us introduce[3,9] the local rate, $\lambda(t)$, defined by:

$$\lambda(t) = \lim_{\Delta t \to 0} \frac{E\{N(t,t + \Delta t)\}}{\Delta t} \tag{2.20}$$

$\lambda(t)$ is thus the local number of events per unit time. In general for a nonstationary process, the local rate is a function of time. The counts PSD function, $S_c(w)$, of a stationary process ($\lambda(t) = \lambda$) is given by:[3]

$$S_c(w) = \frac{\lambda}{2\Pi}\left(1 + \int_{-\infty}^{\infty} (h(\tau) - \lambda)\exp(-jw\tau)d\tau\right) \tag{2.21}$$

where $h(\tau)$ is the conditional intensity function given in Equation 2.9 $S_c(w)$ is the Fourier transform of the counts autocovariance. Methods for estimating the PSD function have been reported in the literature.[9,11]

IV. SOME COMMONLY USED MODELS

A. Introduction

The event generating process is usually to be estimated, or modeled, with the aid of the finite time observed data. The various models are given in terms of the probability distribution functions. The motivation for modeling the point process is mainly to represent the event generating process in a concise parametric form. This allows the detection of changes in the process (due to pathology, for example) and comparison of samples from various processes.

In a stationary point process, the underlying probability distributions do not vary with time. Hence phenomena, common in biological signals, such as fatigue and adaptation, produce nonstationarities. Testing stationarity and detecting trends are important steps in the investigation of the point process; in fact, the initial step of analysis must be the testing of the validity of the stationarity hypothesis. In the remainder of this section, various distribution models will be discussed. These models have been used extensively for modeling neural spike trains, EMG, R-R intervals, and other biological signals.

B. Renewal Processes

An important class of point processes often used in modeling biological signals is the class of renewal processes. Renewal processes are processes in which the intervals between events are independently distributed with identical probability distribution function, say $g(t)$.

In neural modeling it is commonly assumed[8] that the spike occurrences are of a regenerative type which means that the spike train is assumed to be a renewal process. This is used, however, only in cases of spontaneous activity. In the stimulated spike train, the neuron reacts and adapts to the stimuli so that the interval independency is violated.

Consider the intensity function, $h(t)$ (Equation 2.9), of the renewal process. Recall that $h(t)\Delta t$ is the probability of an event occurring in the interval $(t + \Delta t)$ given that an event has occurred at $t = 0$. The event can be the first, second, third, etc. occurrence during the time interval $(0,t)$.

It can be shown[14] that when k events have occurred during the interval $(0,t)$, the intensity function of the renewal process becomes:

$$\begin{aligned} h(t) &= g(t) + [g(t) * g(t)] + [g(t) * g(t) * g(t)] \\ &\quad + [g(t) *,\ldots, *g(t)] \end{aligned} \tag{2.22}$$

where (*) denotes convolution and the last term contains $(k - 1)$ convolutions. Equation 2.22 is better represented via the Laplace transformation. Define

$$H(s) = L[h(t)]$$
$$G(s) = L[g(t)] \tag{2.23}$$

Then we get from Equation 2.22:

$$H(s) = \sum_{i=1}^{k} (G(s))^i = \frac{G(s)(1 - G^k(s))}{1 - G(s)} \tag{2.24}$$

Hence, to characterize an ordinary renewal process, all that is required is an estimation of the parameters of the PDF, g(t), possible by means of a histogram.

When assuming a renewal process, it is first required to test the interval independence hypothesis. Several tests have been suggested, e.g., References 1 and 15; a few of these will be briefly discussed here.

1. Serial Correlogram

The assumption of interval independency (in the sense of weak stationarity) can be tested using the estimation of the serial correlation coefficients defined in Equation 2.17 and estimated by Equation 2.18. The exact distribution of $\hat{\rho}_k$ is, of course, unknown. However, under the assumption that the process is a renewal process and for sufficiently large N, the random variable $\hat{\rho}_k/(n - 1)^{1/2}$ $(k > 0)$ has approximately normal distribution,[15] with zero mean and unit variance. The null hypothesis H_0 is that the interval sequence $\{T_1, T_2, \ldots, T_N\}$ is drawn from a renewal process. The alternative, H_1, hypothesis is that the intervals are identically distributed, but are not independent.

A test based on $\hat{\rho}_k$ will be to reject the renewal hypothesis H_0, if:

$$\frac{|\hat{\rho}_k|}{(n - 1)^{1/2}} > z_{\alpha/2} \tag{2.25}$$

where α is a predetermined significance level and $z_{\alpha/2}$ is given by the integral equation over the normalized (0,1) gaussian distribution:

$$\frac{1}{(2\Pi)^{1/2}} \int_{-\infty}^{z_{\alpha/2}} \exp(-z^2/2)dz = 1 - \alpha/2 \tag{2.26}$$

(e.g., see Bendat and Piersol,[27] Chapter 4). It has been argued that measurement errors (in the case of neural spike trains)[14,28] may introduce trends and dependencies between intervals, thus rendering the serial correlogram test unreliable.

Perkel et al.[13] have suggested subjecting the sequence of intervals to random shuffling and recomputing the correlation coefficients. Serial correlation due to the process (if it exists) will be destroyed by the random shuffling. Computational errors, however, exist in the estimation of both original and shuffled correlations. A test for independence can then be constructed from the comparison of the two correlograms (e.g., by means of the sum of squares of the difference between corresponding correlation coefficients). Other tests have been suggested.[35]

2. Flatness of Spectrum

A renewal process has a flat intervals PSD function. Deviations from a flat spectrum can be used as a test for interval independence.[1] When the spectrum is estimated by the periodogram (Chapter 8, Volume I), the flatness can be tested by the quantities C_j:

$$C_j = \frac{\sum_{i=1}^{j} c_i}{\sum_{i=1}^{N/2-1} c_i}$$

$$j = 1,2,\ldots,N/2 - 1 \qquad (2.27)$$

where c_i, $i = 1,2, \ldots, N/2 - 1$, are the elements of the periodogram. The quantities C_j of Equation 2.27 represent the order statistics[29] from a uniform distribution. The Kolmogorov-Smirnov statistics[29] may be used to test the C_j's.

3. A Nonparametric Trend Test

The renewal process is characterized by the equal distribution of intervals, in addition to interval independence. A common source for "nonrenewalness" is the presence of a long-term trend in the data. Several tests to detect the presence of trends have been suggested. Let $T_1,T_2, \ldots ,T_N$ be N observations of the intervals, and let $G_j(T)$ be the cumulative distribution function of the *jth* observation. Define the null hypothesis, H_0, as the one for which $G_i(T) = G_j(T)$ for $i,j = 1,2, \ldots ,N$ and all T_j are statistically independent.

Suppose now that the process is not a renewal one and a positive trend exists. An alternative hypothesis, H_2, can be defined as $G_1(T) \geqslant G_2(T), \ldots \geqslant G_n(T)$ for all T and at least one of the inequalities holds. A test statistic, D, known as the Mann-Whitney statistic[26] can be used:

$$D = \Sigma_{i<j} \Sigma(j - i)u_{ij} \qquad (2.28A)$$

where

$$u_{i,j} = \begin{cases} 1 & T_i \leqslant T_j \\ 0 & T_i > T_j \end{cases} \qquad (2.28B)$$

If the intervals between events tend to increase with time, the T's will increase with their subscripts, causing the statistic D to be large. It can be shown that for sufficiently large n, given H_0 is true, D is approximately normally distributed with:

$$E\{D|H_0\} = \frac{n^3 - n}{12}$$

$$\mathrm{Var}\{D|H_0\} = \frac{n^2(n + 1)^2(n - 1)}{36} \qquad (2.29)$$

The test, therefore, calls for rejecting the H_0 assumption of no trend (hence the necessary requirement for renewal process) for larger values of D.

C. Poisson Processes

Poisson processes are a special case of renewal processes, in which the identical interval distribution is a Poisson distribution. In the theory of point processes, the Poisson process, due to its simplicity, plays a somewhat analogous role to that of normal distribution in the study of random variables.

The Poisson process, with rate λ, is defined by the requirement that for all t, the following exists as $\Delta t \rightarrow 0$:

$$\text{Prob}\{N(t,t + \Delta t) = 1\} = \lambda\Delta t + 0(\Delta t) \tag{2.30}$$

The constant rate, λ, denotes the average events per unit time. An important aspect of the definition (Equation 2.30) is that the probability does not depend on time. The probability of having an event in $(t,t + \Delta t)$ does not depend on the past at all.

It is well known that for a random variable for which Equation 2.30 holds, the probability of r events occurring in t (starting from some arbitrary time origin) is

$$\text{Prob}(N(t) = r) = \frac{(\lambda t)^r}{r!} \exp(-\lambda t) \tag{2.31}$$

which is the Poisson probability. The probability of having zero events in time T, followed by one event in the interval T + dt, is given by the joint probability of the two. However, the two probabilities are independent, due to the nature of the Poisson process. Also the probability of having one event in the interval T + dt is by Equation 2.30 λdt, hence,

$$p(T)dt = \frac{(\lambda T)^0 \exp(-\lambda T)}{0!} \cdot \lambda dt \tag{2.32A}$$

or

$$p(T) = \lambda \exp(-\lambda T) \tag{2.32B}$$

Equation 2.32B gives the PDF of the Poisson distribution.

Refer to Figure 4 and consider the *nth* order intervals, $T_i^{(n)}$, given by Equation 2.1. Due to the overlapping, the *nth* order interval will no longer be Poisson distributed. Consider the occurrence of n − 1 events before time t; the probability for this is (from Equation 2.31):

$$\text{Prob}(N(t) = n - 1) = \frac{(\lambda t)^{n-1}}{(n - 1)!} \exp(-\lambda t) \tag{2.33}$$

The probability that in the following time interval of t + dt one and only one event will occur is λdt. Since the two are independent, the joint probability of their occurrence is given by:

$$p_j(t)dt = \frac{(\lambda t)^{n-1}}{(n - 1)!} \lambda \exp(-\lambda t)dt \tag{2.34A}$$

or

$$p(T^{(n)}) = \frac{\lambda^n(T^{(n)})^{n-1}}{(n - 1)!} \exp(-\lambda t) \tag{2.34B}$$

The PDF of the *nth* order interval given by Equation 2.34B is known as the Gamma distribution.

The survivor function (Equation 2.6) for the Poisson process is given by integrating Equation 2.32B:

$$R_i(T) = \text{Prob}(T_i > T) = \exp(-\lambda T) \tag{2.35}$$

Consider now the autocovariance and the spectrum of the Poisson process. Since the interval, T_i, is independent of T_j for all $i \neq j$, the autocovariance of the process (Equation 2.15) becomes a delta function. Its Fourier transform, the interval power spectral density function (Equation 2.18), is thus constant (flat):

$$S_I(s) = \frac{1}{2\Pi\lambda^2} \tag{2.36}$$

It can also be shown that for the Poisson process the relative intensity function $h(\tau) = \lambda$. Hence the counts' power spectral density function (Equation 2.20) is also flat, with:

$$S_c(w) = \frac{\lambda}{2\Pi} \tag{2.37}$$

Several statistics to test the hypothesis that a given sequence of intervals was drawn from a Poisson process have been suggested. For the Poisson process, the quantities

$$P_i = t_i/t_N$$
$$i = 1,2,\ldots,N \tag{2.38}$$

(Figure 4) represent the order statistics from a random sample size N, which is uniformly distributed with zero mean and unit variance.[14] A modification to Equation 2.38 shows[1] that when rearranging the intervals sequence to generate a new sequence $\{T_i^0\}$ in which $T_{i+1}^0 \geqslant T_i^0$ the quantities

$$\tilde{P}_i = \frac{1}{t_N} \sum_{j=1}^{N+2-i} T_j^0$$
$$i = 1,2,\ldots,N \tag{2.39}$$

also represent a similar order statistics. The Kolmogrov-Smirnov[1,29] statistics can then be used to test the Poisson hypothesis.

Other tests based, for example, on the coefficient of variations[15] have been suggested. It is sometimes of interest to test whether the Poisson process under investigation is a homogeneous or nonhomogeneous Poisson process. A nonhomogeneous Poisson process is one in which the rate of occurrence, λ, is not constant but time dependent — in other words, a Poisson process with trend in the rate of occurrence. The Wald-Wolfowitz run test[27] may be used for this task. For this test we define a set of equal arbitrary time interval lengths (TIL). If the number of events in the TIL exceeds the expected number for this interval, a (+) sign is attached to the TIL. If the number of events is below the expected number, a (−) sign is attached. When the number of events equals the expected number, the TIL is discarded. A sequence of (+) and (−) signs is thus generated. The number of runs, r, is determined by summing up each uninterrupted sequence of (+) or (−). The sequence (+ + − − − − + − + +) yields r = 5.

The total number of (+) is denoted by N_+ and the total number of (−) is denoted by N_-. The mean and variance of r are

$$\hat{\mu}_r = 1 + (2N_+N_-)/N \tag{2.40A}$$

and

$$\hat{\sigma}_r^2 = (\hat{\mu}_r - 1)(\hat{\mu}_r - 2)/(N + 1) \tag{2.40B}$$

For large samples, the approximate standard normal variate,

$$z = (|r - \hat{\mu}_r| - 0.5)/\hat{\sigma}_r \tag{2.41}$$

is used to test the data.[27]

D. Other Distributions

In some cases the process under investigation does not fit the simple Poisson distribution. Other distributions have been found useful in describing biological point processes. The more commonly used ones are discussed here.

1. The Weibull Distribution

This distribution is sometimes used to model renewal neural spike trains.[13] The probability density function of the Weibull distribution is[30,31]

$$p(T,\upsilon,\epsilon,k) = \begin{cases} \dfrac{k}{\upsilon - \epsilon}\left(\dfrac{T - \epsilon}{\upsilon - \epsilon}\right)^{k-1} \exp\left(-\left(\dfrac{T - \epsilon}{\upsilon - \epsilon}\right)^{k}\right) & ; \quad T > \epsilon \\ 0 & ; \quad T < \epsilon \end{cases}$$

$$k > 0 \quad ; \quad \upsilon > \epsilon \tag{2.42}$$

and the cumulative distribution function:

$$P(T,\upsilon,\epsilon,k) = \begin{cases} 1 - \exp\left(-\left(\dfrac{T - \epsilon}{\upsilon - \epsilon}\right)^{k}\right) & ; \quad T > \epsilon \\ 0 & ; \quad T < \epsilon \end{cases}$$

$$k > 1 \quad ; \quad \upsilon \geqslant \epsilon \tag{2.43}$$

A random variable, T, with a Weibull distribution has the expectation and variance given by:[29]

$$E\{T\} = (\upsilon - \epsilon)\Gamma(1 + k^{-1}) \tag{2.44A}$$

$$\mathrm{Var}\{T\} = (\upsilon - \epsilon)^2[\Gamma(1 + 2k^{-1}) - \Gamma^2(1 + k^{-1})] \tag{2.44B}$$

where $\Gamma(\cdot)$ is the gamma function. Note that for $k = 1$ the Weibull density reduces to the exponential density.

2. The Erlang (Gamma) Distribution

This distribution has also been used to model renewal neural spike trains.[13] Its probability density function is

$$p(T,k,\lambda) = \begin{cases} \frac{\lambda}{\Gamma(k)} (\lambda T)^{k-1} \exp(-\lambda T) & ; \quad T > 0 \\ \\ 0 & ; \quad T < 0 \end{cases} \qquad k > 0 \tag{2.45}$$

where $\Gamma(\cdot)$ is gamma function. A random variable, T, with Erlang distribution has the expectation and variance:

$$E\{T\} = \frac{r}{\lambda}$$

$$\mathrm{Var}[T] = \frac{r}{\lambda^2} \tag{2.46}$$

The Erlang distribution with $r = 1$ becomes the exponential distribution.

3. Exponential Autoregressive Moving Average (EARMA)

An important stationary class of point processes which sometimes is useful as an alternative to the Poisson process is the ARMA processes. In these nonrenewal processes, exponentially distributed intervals are statistically dependent upon past intervals in an ARMA sense (Chapter 7, Volume I). Such a process is termed EARMA (p,q) where p is the AR order and q is the MA order of the process. The general EARMA process is called exponential autoregressive ($EAR_{(p)}$) process when $q = 0$ and is called exponential moving average ($EMA_{(q)}$) process when $p = 0$. For a more detailed discussion on these processes, see Cox and Isham.[3]

4. Semi-Markov Processes

A sequence of random variables, x_n, is called Markov if for any n we have the conditional probability:

$$P(x_n|x_{n-1},x_{n-2},\ldots,x_1) = P(x_n|x_{n-1}) \tag{2.47}$$

namely, the probability of the current event depends only on the event preceding it. Assume now that the random variable, x_n, is a discrete random variable taking the values $a_1,a_2, \ldots ,a_n$. The sequence $\{x_n\}$ is then called a Markov chain. A semi-Markov process is a process in which the intervals are random-variably drawn from any one of a given set of distribution functions.[3] The switching between one probability function to another is controlled by a Markov chain.

Consider the case with k "classes" or "types" and a set of k^2 distribution functions $F_{i,j}$, $i,j = 1,2, \ldots ,k$. Assume that each interval of the point process is assigned a "class" type, $1,2, \ldots ,k$. The assignment is determined by a Markov chain with transition matrix $P = (P_{i,j})$. A sequence of intervals beginning with type i and ending with type j is drawn from the distribution $F_{i,j}$. The transition matrix P is such that when an interval had been assigned a class i, the probability of the next interval to get the class j is $P_{i,j}$.

A special case of the Semi-Markov process is the two-State Semi-Markov model (TSSM). Here the transition matrix P is

$$P = \begin{bmatrix} P_1 & 1 - P_1 \\ 1 - P_2 & P_2 \end{bmatrix} \tag{2.48}$$

and the four distributions $F_{i,j}$ are

$$\begin{aligned} F_{1,1} &= F_{1,2} = F_1 \\ F_{2,1} &= F_{2,2} = F_2 \end{aligned} \tag{2.49}$$

Equation 2.49 states that the interval probability distribution depends only on the type of the interval and not on adjacent types. In a Semi-Markov process for which Equation 2.49 holds, the number of consecutive intervals which have the same distribution is geometrically distributed.[3]

The TSSM model has been applied to spike train activity analysis. It was found, however,[11] that it can be used only for a limited part of all experimental stationary data. The more general nongeometric Semi-Markov model has been applied with more success.[32] The nongeometric two-state Semi-Markov process has also been applied to neural analysis by De Kwaadsteniet.[11] The term semialternating renewal (SAR) is used there.

V. MULTIVARIATE POINT PROCESSES

A. Introduction

In multivariate point processes, two or more types of points are observed. This may be the case, for example, when two or more univariate point processes are investigated and the relationship or dependence between them is sought. Another example may be where several different processes are recorded together. Multispike trains are common in neurophysiology[10,33] when an electrode picks the spikes of the neuron under investigation together with spikes from neighboring neurons. It is often possible to distinguish between the various neuron spikes based on the difference in pulse shapes.[10] The record thus can be considered a multivariate point process. A similar process occurs when recording muscle activity. Several motor units form a multivariate point process.

In general, the various types of the multivariate process are dependent on one another; it is therefore necessary to have the conditional probability functions in order to characterize the process.

B. Characterization of Multivariate Point Processes

We shall discuss the stationary case where just two types of events are present in the multivariate process.[3] The general case can easily be extended from the following discussion.

Let ${}^1N(t)$ and ${}^2N(t)$ be the number of type 1 events and type 2 events, in the interval (0,t), respectively. We shall retain the orderliness of the process (in the sense of Equation 2.2). Hence, no simultaneous events are allowed. It is often required, however, to assume that events of different types may occur simultaneously, as is the case in multispike neural train analysis. We can easily overcome the problem by assigning a separate class to all events in which two types have occurred simultaneously.

Denote the intervals ${}^1T(i)$ and ${}^2T(i)$ as the intervals between the *ith* and (i − 1)*th* events of type 1 and type 2, respectively. The relationships between the counts ${}^1N(t)$ and ${}^2N(t)$ and the intervals are similar to those of the univariate process (Equation 2.11), namely,

$$^1N(t_1) < n_1$$

$$^2N(t_2) < n_2 \tag{2.50A}$$

if and only if

$$\sum_{i=1}^{n_1} {}^1T(i) > t_1$$

$$\sum_{i=1}^{n_2} {}^2T(i) > t_2 \tag{2.50B}$$

Similar to the univariate process we shall define the cross intensity function $^1_2h(\tau)$ as a generalization of Equation 2.9.

$$^1_2h(\tau) = \lim_{\Delta t \to 0} \left(\frac{\text{Prob}\{^1N(t + \tau, t + \tau + \Delta t) > 0 | ^2N(t,t + \Delta t) > 0\}}{\Delta t} \right) \tag{2.51}$$

The cross intensity function, $^2_1h(\tau)$, is similarly defined. The cross intensity, $^1_2h(\tau)\Delta t$, yields the probability of having an event of type 1 at time τ, given an event of type 2 has occurred at the origin. Note that $^1_1h(\tau)$ is the univariate conditional intensity. The complete intensity function of the multivariate process is defined (see Equation 2.8):

$$^1h_0(\tau) = \lim_{\Delta t \to 0} \left(\frac{\text{Prob}\{^1N(\tau,\tau + \Delta t) > 0, ^2N(\tau,\tau + \Delta \tau) = 0\}}{\Delta t} \right)$$

$$^2h_0(\tau) = \lim_{\Delta t \to 0} \left(\frac{\text{Prob}\{^1N(\tau,\tau + \Delta \tau) = 0, ^2N(\tau,\tau + \Delta t) > 0\}}{\Delta t} \right) \tag{2.52}$$

and the complete intensity function of simultaneous occurrence of the two types at time τ is

$$^{1,2}h_0(\tau) = \lim_{\Delta t \to 0} \left(\frac{\text{Prob}\{^1N(\tau,\tau + \Delta \tau) > 0, ^2N(\tau,\tau + \Delta t) > 0\}}{\Delta t} \right) \tag{2.53}$$

It is sometimes required that one ignore the different types of the multivariate process and consider it a univariate process. The conditional intensity function of such a process is given, in terms of the intensities of the multivariate process:[3]

$$h(\tau) = \frac{\lambda_1}{\lambda_1 + \lambda_2} (^1_1h(\tau) + {}^2_1h(\tau)) + \frac{\lambda_2}{\lambda_1 + \lambda_2} (^1_2h(\tau) + {}^2_2h(\tau)) \tag{2.54}$$

where λ_1, λ_2 are the rates of the first and second types.

Covariance and spectral analysis of multivariate processes are important tools in the investigation of these processes: they have been applied[33] to the analysis of neurological signals. Define the cross covariance density,[3] $^1_2C(\tau)$,

$$^1_2C(\tau) = \lim_{\Delta t \to 0} \left(\frac{\text{Cov}\{^1N(t + \tau,t + \tau + \Delta t), ^2N(t,t + \Delta t)\}}{\Delta t} \right) = \lambda_2 \, {}^1_2h(\tau) - \lambda_1\lambda_2 \tag{2.55}$$

with the cross covariance density, ${}^{2}_{1}C(\tau)$, defined similarly. The cross spectral density function, ${}^{1}_{2}S(w)$, is the Fourier transform of Equation 2.55:

$$ {}^{1}_{2}S(w) = \frac{1}{2\Pi} \int_{-\infty}^{\infty} {}^{1}_{2}C(\tau)\exp(-jw\tau)d\tau \tag{2.56} $$

with the cross spectral density, ${}^{2}_{1}S(w)$ defined similarly. A discussion concerning the application of the cross spectral density function to neural spikes processing has been given by Glaser and Ruchkin.[33]

C. Marked Processes

A multivariate point process can be expressed as a univariate process with a marker attached to each point marking its type. Such is often the case in neurophysiological recordings of action potentials. A microelectrode may record the action potentials generated by several neurons in its vicinity. Action potentials (spikes) of the various neurons, as recorded by the microelectrode, differ from one another in shape and can be classified[10] by means of wavelet detection methods (see also Chapter 1, Volume II). The recording of the microelectrode, known as multispike train,[10] can thus be analyzed as a marked point process.

We shall denote the mark of the *ith* point by M_i and the accumulated mark at time t by M(t), hence,

$$ M(t) = \Sigma M_i \tag{2.57} $$

where the summation takes place over all points in the interval (0,t). M(t) is a random variable with statistics to be estimated from the given data.

In the general case we shall be interested in the joint probability distribution of (M(t),N(t)). From it we can derive the dependence (if any) of the different types on one another. Consider the more simple case where the point process has a rate, λ, and the marks are independent (of one another and of the point process) and are uniformally distributed. Assume a record of length N with N(A) = n and M(A), the sum of the n independent marks. It can be shown[3] that:

$$ \begin{aligned} E\{M(A)\} &= \lambda|A|E\{M\} \\ Var\{M(A)\} &= \lambda|A|Var\{M\} + (E\{M\} + (E\{M\})^2 Var\{N(A)\} \\ Cov\{M(A),N(A)\} &= E\{M\}Var\{N(A)\} \end{aligned} \tag{2.58} $$

Methods for the analysis of the marked point process with dependencies between marks and process have been reported in the literature; these are, however, outside the scope of this book.

REFERENCES

1. **Cox, D. R. and Lewis, P. A. W.,** *The Statistical Analysis of Series of Events,* Methuen, London, 1966.
2. **Lewis, P. A. W., Ed.,** *Stochastic Point Processes: Statistical Theory and Applications,* Wiley-Interscience, New York, 1972.
3. **Cox, D. R. and Isham, V.,** *Point Processes,* Chapman and Hall, London, 1980.
4. **Brillinger, D. R.,** Comparative aspects of the study ordinary time series and of point processes, in *Developments in Statistics,* Krishnaiah, P. R., Ed., Academic Press, New York, 1978, 33.

5. **Sayers, B. McA.,** Inferring significance from biological signals, in *Biomedical Engineering Systems,* Clynes, M. and Milsum, J. H., Eds., McGraw-Hill, New York, 1970, chap. 4.
6. **Anderson, D. J. and Correia, M. J.,** The detection and analysis of point processes in biological signals, *Proc. IEEE,* 65(5), 773, 1977.
7. **Ten Hoopen, M. and Penver, H. A.,** Analysis of sequences of events with random displacements applied to biological systems, *Math. Biosci.,* 1, 599, 1967.
8. **Fienberg, S. E.,** Stochastic models for single neuron firing trains: a survey, *Biometrics,* 30, 399, 1974.
9. **Lago, P. J. and Jones, N. B.,** A note on the spectral analysis of neural spike train, *Med. Biol. Eng. Comput.,* 20, 44, 1982.
10. **Abeles, M. and Goldstein, M. H.,** Multispike train analysis, *Proc. IEEE,* 65(5), 762, 1977.
11. **De Kwaadsteniet, J. W.,** Statistical analysis and stochastic modeling of neural spike train activity, *Math. Biosci.,* 60, 17, 1982.
12. **Ten Hoopen, M.,** The correlation operator and pulse trains, *Med. Biol. Eng.,* 8, 187, 1970.
13. **Perkel, D. K., Gerstein, G. L., and Moore, G. P.,** Neuronal spike trains and stochastic point processes. I. The single spike train. II. Simultaneous spike trains, *Biophys. J.,* 7, 391, 419, 1967.
14. **Landolt, J. P. and Correia, M. J.,** Neuromathematical concepts of point processes theory, *IEEE Trans. Biol. Med. Eng.,* 25(1), 1, 1978.
15. **Yang, G. L. and Chen, T. C.,** On statistical methods in neuronal spike train analysis, *Math. Biosci.,* 38, 1, 1978.
16. **Sampath, G. and Srinivasan, S. K.,** *Stochastic Models for Spike Trains of Single Neurons,* Lecture Notes in Biometrics, Vol. 16, Springer-Verlag, Berlin, 1977.
17. **Clamann, H. P.,** Statistical analysis of motor unit firing patterns in human skeletal muscle, *Biophys. J.,* 9, 1233, 1969.
18. **Parker, P. A. and Scott, R. N.,** Statistics of the myoeletric signal from monopolar and bipolar electrodes, *Med. Biol. Eng.,* 11, 591, 1973.
19. **Lago, P. J. A. and Jones, N. B.,** Turning points spectral analysis of the interference myolectric activity, *Med. Biol. Eng. Comput.,* 21, 333, 1983.
20. **Andreassen, S.,** Computerized analysis of motor unit firing, in *Progress in Clinical Neurophysiology, Vol. 10, Computer Aided Electromyography,* Desmedt, J. E., Ed., S. Karger, Basel, 1983, 150.
21. **Ten Hoopen, M.,** R-wave sequences treated as a point process, progress report 3, *Inst. Med. Phys.,* TNO, Utrecht, Netherlands, 1972, 124.
22. **Goldstein, R. E. and Barnett, G. O.,** A statistical study of the ventricular irregularity of atrial fibrillation, *Comput. Biomed. Res.,* 1, 146, 1967.
23. **Schafer, R. W. and Markel, J. D., Eds.,** *Speech Analysis,* IEEE Press, New York, 1979.
24. **Kasuya, H., Kobayashi, Y., Kobayashi, T., and Ebihara, S.,** Characterization of pitch period and amplitude perturbation in pathological voice, in *Proc. IEEE Int. Conf. Acoust. Speech Signal Process.,* IEEE, New York, 1983, 1372.
25. **Brassard, J. R., Correia, M. J., and Landolt, J. P.,** A computer program for graphical and iterative fitting of probability density functions to biological data, *Comput. Prog. Biomed.,* 5, 11, 1975.
26. **Lehmann, E.,** *Non Parametrics: Statistical Methods Based on Ranks,* Holden-Day, San Francisco, 1975.
27. **Bendat, J. S. and Piersol, A. G.,** *Random Data: Analysis and Measurement Procedures,* Wiley-Interscience, New York, 1971.
28. **Shiavi, R. and Negin, M.,** The effect of measurement errors on correlation estimates in spike interval sequences, *IEEE Trans. Biomed. Eng.,* 20, 374, 1973.
29. **Mood, A. M., Graybill, F. A., and Boes, D. C.,** *Introduction to the Theory of Statistics,* 3rd ed., McGraw-Hill, Kogakusha, Tokyo, 1974.
30. **Parzan, E.,** *Stochastic Processes,* Holden-Day, San Francisco, 1962.
31. **Mann, N. R., Schafer, R. E., and Singpurwalla, N. S.,** *Methods for Statistical Analysis of Reliability and Life Data,* Wiley-Interscience, New York, 1974.
32. **Ekholm, A.,** A generalization of the two state two interval semi-Markov model, in *Stochastic Point Processes,* Lewis, P. A., Ed., Wiley-Interscience, New York, 1972.
33. **Glaser, E. M. and Ruchkin, D. S.,** *Principles of Neurobiological Signal Analysis,* Academic Press, New York, 1976.
34. **Bartlett, M. S.,** The spectral analysis of point processes, *J. R. Stat. Soc. Ser. B,* 25, 264, 1963.
35. **Li, H. F. and Chan, F. H. Y.,** Microprocessor based spike train analysis, *Comput. Prog. Biomed.,* 13, 61, 1981.

Chapter 3

SIGNAL CLASSIFICATION AND RECOGNITION

I. INTRODUCTION

Modern biomedical signal processing requires the handling of large quantities of data. In the neurological clinic, for example, routine examinations of electroencephalograms are usually performed with eight or more channels, each lasting several tens of seconds. In more elaborate examinations for sleep disorders analysis, hours-long records may be taken. Several hours of electrocardiographic recordings are sometimes required from patients recovering from heart surgery. Various screening programs are faced with the problem of handling a large number of short-term ECG and other signals.

Storing and analyzing such large quantities of information have become a severe problem. In some cases manual analysis is cost prohibitive; in others, it is completely impossible. The problem has therefore been recognized as an important part of any modern signal and pattern analysis system. Signals are in essence one-dimensional patterns. The methods and algorithms developed for pattern recognition are in general applicable to signal analysis.

The topics discussed in this chapter are based on the decision — theoretic approach to pattern recognition. A different approach — the syntactic method — is discussed in Chapter 4.

The signal to be analyzed, stored, or transmitted contains quite often some redundancies. This may be due to some built-in redundancies, added noise, or the fact that the application at hand does not require all the information carried by the signal. The first step for sophisticated processing will be that of data compression. Irrelevant information is taken out such that the signal can be represented more effectively. One accepted method for data compression is by features extraction (refer to Figure 2, Chapter 1, Volume I). Based on some *a priori* knowledge about the signal, features are defined and extracted. The signal is then discarded, with the features being its representation. Features must thus be carefully selected. They must contain most of the relevant information while most of the redundancies are discarded. Optimal feature selection routines are available. For some applications compression is required for storage or transmition purposes, so that the signal must, at a later stage, be reconstructed. Features based on time series analysis (ARMA, AR — see Chapter 7, Volume I) can be used for such applications where the reconstructed signal has the same spectral characteristics as the original one. In other applications, automatic classification is required. Compression is then performed with features that not necessarily allow reconstruction, but provide distinction between classes.

Any linear or nonlinear transformation of the original measurement can be considered as features, provided they allow reconstruction or give discriminant power. Various transformations, optimal in some sense, have been used to compress signal data. These transformations can be used without the need for *a priori* knowledge of the signal. Other features require some assumptions on signal properties — these may be, for example, the order of the ARMA model or the range of allowable peak of a waveform.

In many cases, the features extracted are statistically dependent on one another; some methods, however, provide independent features. The computational cost and time required for the feature extraction process usually dictate the need to reduce the number of features as much as possible. A compromise has to be taken between that demand and the accuracy (in reconstruction or classification) requirement. Methods for (sub) optimal determination of the number of features are available; some are discussed in this chapter.

The material covered in this chapter is based on the vast literature on pattern and signal recognition textbooks,[1-5] reference books,[6,7] and papers.[8] Signal classification methods have

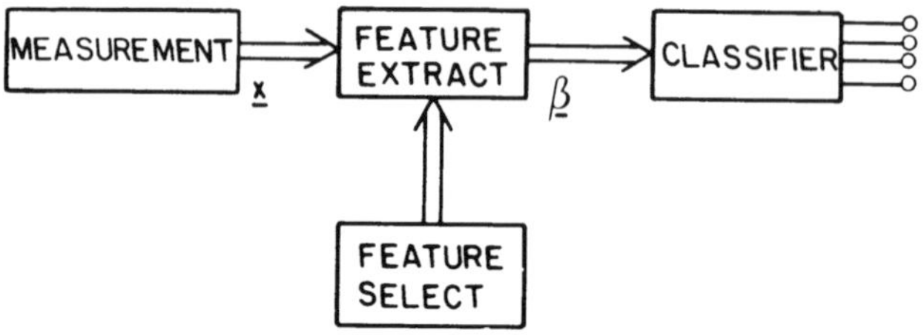

FIGURE 1. Schematic representation of feature extraction, selection, and classification system.

been applied to various biomedical problems such as neural signals,[9] electroencephalography,[10-18,53] electrocardiography,[19-32,53] phonocardiography,[33] breath sounds,[34,35] electromyography,[36] voice,[37,38] and many more.

The recognition of the signal is done by means of a classification process. The machine has *a priori* knowledge on the types (classes) of signals under consideration. An unknown signal is then classified into one of the known classes. One of the classes can be a "reject" or "unknown" class. The input to the recognition machine is a set of N measurements given in terms of the vector, $\underline{x}$, (e.g., samples of the signal) and called the "pattern vector". The pattern vector contains all the measured information available about the signal. A set of features, arranged as the features vector, $\underline{\beta}$, is extracted from the pattern vector (Figure 1). The classifier operates on the features vector $\underline{\beta}$, with functions known as the "decisions" or "discriminant functions" to perform the classification. The decision functions, $d_i(\underline{\beta})$, are scalars and are single-valued functions of $\underline{\beta}$. In general when M classes are involved, we use M decision functions, $d_i(\underline{\beta})$, $i = 1,2,\ldots,M$. The signal with pattern vector, x, and feature vector, $\underline{\beta}$, will be classified into the *ith* class, w_i, if:

$$d_i(\underline{\beta}) > d_j(\underline{\beta});\ i,j = 1,2,\ldots,M \qquad i \neq j$$

Consider the simple example depicted in Figure 2. Here the features vector is of dimension three $\underline{\beta}^T = [\beta_1,\beta_2,\beta_3]$ and two classes, w_1, w_2, are given. The two clusters of features of w_1 and w_2 belong to signals known in advance to be in either w_1 or w_2. These are known as the training set. The clusters of features may be considered as an estimate for the probability distribution of the features. The projections of the clusters in the three-dimensional feature space are shown in Figure 2. It is clearly seen that classification of the two classes can be made with feature β_2 alone, since the projections of w_2 and w_1 ($w_2^{(2)}$ and $w_1^{(2)}$) do not overlap. The projections on other feature-axes show that overlapping exists between the two classes. A linear decision function can be drawn in the (β_1,β_2) or (β_2,β_3) planes to discriminate between the two classes.

An example for the procedure described above can be that of automatic classification of ECG signals. Here samples of records of ECGs of normal and pathological states are given. These have been diagnosed manually by the physician and constitute the training set. From this given set, templates (for the normal state and each one of the pathological states) are generated and the statistics of each class are estimated. It is clear that the more information there is in the training set, the better is the training process and the probability of correct classification.

In some cases training sets are not *a priori* classified. The system must then "train" itself by means of unsupervised learning. Cluster-seeking algorithms have to be used in order to automatically identify groups that can be considered classes. Unsupervised recognition sys-

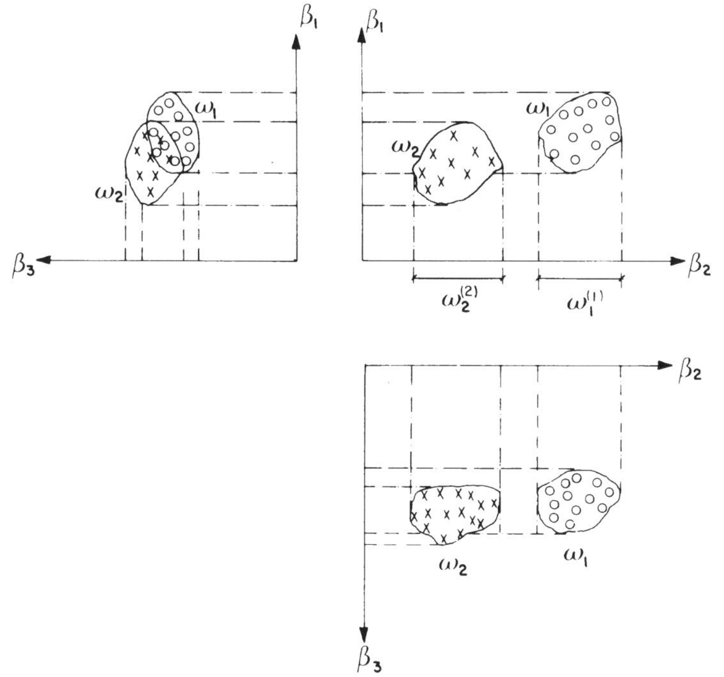

FIGURE 2. Projections of clusters in the features space.

tems require a great deal of intuition and experimentation. The interested reader is referred to the pattern recognition literature.

Two important topics are discussed in this chapter: features selection and signal classification. Both were included in one chapter since in many cases there are similarities in the discussions of the two problems. It was probably logical to open with the discussion on features selection and signal compression since in most cases these are done *prior* to classification (Figure 2). It was found, however, that from the point of view of material presentation it is more convenient to discuss the topic of classification first.

II. STATISTICAL SIGNAL CLASSIFICATION

A. Introduction

We may look at the signal classification problem in probabilistic terms. Assume that we have M classes and an unknown signal to be classified. We define the hypothesis, H_k, that the signal belongs to the w_k class. The problem then becomes a problem of hypothesis testing. In the case of two classes we have the null hypothesis (usually denoted by H_0, but for convenience denoted here by H_1) that the signal belongs to w_1 and the alternative hypothesis, H_2, that the signal belongs to w_2. It is the task of the classifier to accept or reject the null hypothesis. For this we need the probability density functions of the various classes, which are usually *a priori* unknown. The methods for statistical decision learning and classification are discussed in this section.

B. Bayes Decision Theory and Classification

Assume that all the relevant statistics of the problem at hand are given. In particular the probabilities of the classes $P(w_i)$, $i = 1,2, \ldots ,M$, and the conditional probability densities $p(\underline{\beta}/w_i)$ are *a priori* given.

Consider the two classes problem. Assume we have an unknown signal represented by the features, $\underline{\beta}$. The conditional probability that this signal belongs to the *jth* class, $P(w_j|\underline{\beta})$, is given by Bayes rule:

$$P(w_j|\underline{\beta}) = \frac{p(\underline{\beta}|w_j)P(w_j)}{p(\underline{\beta})} \tag{3.1}$$

where $p(\underline{\beta}|w_j)$ is the conditional probability density of getting $\underline{\beta}$ given the class is w_j, $p(\underline{\beta})$ is the probability density function of $\underline{\beta}$, and $P(w_j)$ is the probability of the *jth* class.

Note also that (for the two classes case):

$$p(\underline{\beta}) = \sum_{j=1}^{2} p(\underline{\beta}|w_j)P(w_j) \tag{3.2}$$

Equation 3.1 gives the *a posteriori* probability $P(w_j|\underline{\beta})$ in terms of the *a priori* probability $P(w_j)$. It is logical to classify the signal $\underline{\beta}$ as follows: if $P(w_1|\underline{\beta}) > P(w_2|\underline{\beta})$ we decide $\beta \in w_1$, and if $P(w_2|\underline{\beta}) > P(w_1|\underline{\beta})$ we decide $\underline{\beta} \in w_2$. If $P(w_1|\underline{\beta}) = P(w_2|\underline{\beta})$ we remain undecided.

Analyzing all possibilities we see that a correct classification occurs when:

$$\underline{\beta} \in w_1 \text{ and } P(w_1|\underline{\beta}) > P(w_2|\underline{\beta})$$

$$\underline{\beta} \in w_2 \text{ and } P(w_1|\underline{\beta}) < P(w_2|\underline{\beta})$$

and an error in classification occurs where:

$$\underline{\beta} \in w_1 \text{ and } P(w_1|\underline{\beta}) < P(w_2|\underline{\beta})$$

$$\underline{\beta} \in w_2 \text{ and } P(w_1|\underline{\beta}) > P(w_2|\underline{\beta})$$

In hypothesis testing language the errors are called the error of the "first kind" and the "second kind" or "false positive" and "false negative". The probability of an error is thus:

$$P(\text{error}|\underline{\beta}) = \begin{cases} P(w_1|\underline{\beta}) \text{ if } P(w_2|\underline{\beta}) > P(w_1|\underline{\beta}) \\ P(w_2|\underline{\beta}) \text{ if } P(w_1|\underline{\beta}) > P(w_2|\underline{\beta}) \end{cases} \tag{3.3}$$

and the average error probability is

$$P(\text{error}) = \int_{-\infty}^{\infty} P(\text{error}|\underline{\beta})p(\underline{\beta})d\underline{\beta} \tag{3.4}$$

It can easily be shown that the intuitive decision rule we have chosen minimizes the average error probability. The Bayes decision rule can be written by means of the conditional probabilities:

$$p(\underline{\beta}|w_1)P(w_1) \gtrless p(\underline{\beta}|w_2)P(w_2) \rightarrow \underline{\beta} \in \begin{cases} w_1 \\ w_2 \end{cases} \tag{3.5}$$

which means that when the left side of the inequality (Equation 3.5) is larger than the right side, we classify $\underline{\beta}$ into w_1; when it is smaller, $\underline{\beta}$ is classified into w_2.

We want now to generalize the decision rule of Equation 3.5. Assume we have M classes. For this case we shall have the probability of $\underline{\beta}$, $p(\underline{\beta})$, given by Equation 3.2, but with summation index running $j = 1, \ldots, M$. We also want to introduce a weight on the various errors. Suppose that when making a classification $\underline{\beta} \in w_i$, we take a certain action, α_i. This may be, for example, the administration of certain medication after classifying the signal as belonging to some illness w_i. We want to attach a certain loss, or punishment, when we take an action α_i when indeed $\underline{\beta} \in w_j$. Denote this loss by $\lambda(\alpha_i|w_j) = \lambda_{ij}$.

Suppose that we observe a signal with features vector $\underline{\beta}$ and consider taking the action α_i. If indeed $\underline{\beta} \in w_j$, we will incur the loss $\lambda(\alpha_i|w_j)$. The expected loss associated with taking the action α_i (also known as the conditional risk) is

$$R(\alpha_i|\underline{\beta}) = \sum_{j=1}^{M} \lambda(\alpha_i|w_j)P(w_j|\underline{\beta}) = \sum_{j=1}^{M} \lambda_{ij}P(w_j|\underline{\beta}) \qquad (3.6)$$

The classification can be formulated as follows. Given a feature vector $\underline{\beta}$, compute all conditional risks $R(\alpha_i|\underline{\beta})$, $i = 1,2, \ldots, M$, and choose the action α_i (classification into w_i) that minimizes the conditional risk (Equation 3.6).

Consider, for example, the two classes case. Equation 3.6 for this case is

$$R(\alpha_1|\underline{\beta}) = \lambda_{11}P(w_1|\underline{\beta}) + \lambda_{12}P(w_2|\underline{\beta})$$

$$R(\alpha_2|\underline{\beta}) = \lambda_{21}P(w_1|\underline{\beta}) + \lambda_{22}P(w_2|\underline{\beta}) \qquad (3.7)$$

Note that λ_{11} and λ_{22} are the loss for correct classification; these are less than the loss for making an error $(\lambda_{12},\lambda_{21})$. We classify $\underline{\beta}$ into w_1 if $R(\alpha_1|\underline{\beta}) < R(\alpha_2|\underline{\beta})$. From Equations 3.7 and 3.1 we get classification into w_1 when:

$$(\lambda_{21} - \lambda_{11})p(\underline{\beta}|w_1)P(w_1) > (\lambda_{12} - \lambda_{22})p(\underline{\beta}|w_2)P(w_2) \qquad (3.8)$$

or

$$\frac{p(\underline{\beta}|w_1)}{p(\underline{\beta}|w_2)} \gtrless \frac{(\lambda_{12} - \lambda_{22})P(w_2)}{(\lambda_{21} - \lambda_{11})P(w_1)} \rightarrow \underline{\beta} \in \begin{cases} w_1 \\ w_2 \end{cases} \qquad (3.9)$$

The left side of the inequality (Equation 3.9) is called the likelihood ratio. The right side can be considered a decision threshold.

In general, classification is performed with discriminant functions. A classification machine computes M discriminant functions, one for each class, and chooses the class yielding the largest discriminant function. The Bayes rule calls for the minimization of Equation 3.6; we can then define the discriminant function of the *ith* class $d_i(\underline{\beta})$ by:

$$d_i(\underline{\beta}) = -R(\alpha_i|\underline{\beta}) \qquad (3.10)$$

and the classification rule becomes: assign the signal with feature vector $\underline{\beta}$ to class w_i if:

$$d_i(\underline{\beta}) > d_j(\underline{\beta}) \text{ for all } j \neq i \qquad (3.11)$$

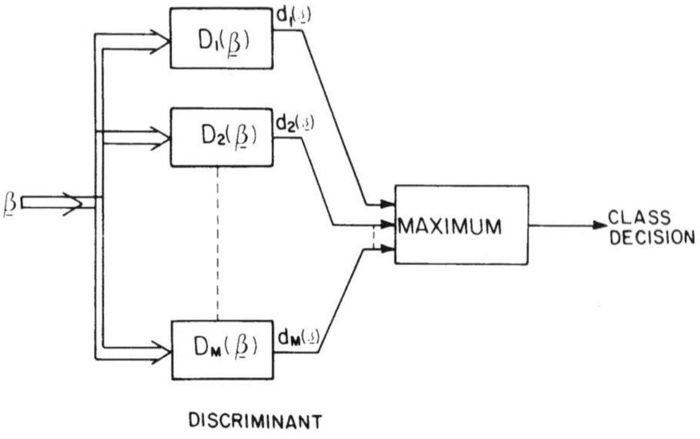

FIGURE 3. Classification by discriminant functions.

Since the logarithm function is a monotonically increasing function, we can take also the logarithm Equation 3.11 without changing the rule. Figure 3 shows the general classifier scheme.

Consider now a simple loss function:

$$\lambda(\alpha_i|w_j) = \begin{cases} 0 & i = j \\ 1 & i \neq j \end{cases}$$

$$i,j = 1,2,\ldots,M \tag{3.12}$$

The average conditional risk of Equation 3.6 becomes:

$$R(\alpha_i|\underline{\beta}) = \sum_{i \neq j} P(w_i|\underline{\beta}) = 1 - P(w_i|\underline{\beta}) \tag{3.13}$$

To minimize the risk, we want to choose that w_i for which $P(w_i|\underline{\beta})$ is maximum. Hence, for this case, known as the "minimum error rate", the classification rule becomes: classify $\underline{\beta}$ into w_i if:

$$P(w_i|\underline{\beta}) > P(w_j|\underline{\beta}) \text{ for all } j \neq i \tag{3.14}$$

We can define the discriminant function as:

$$d_i(\underline{\beta}) = \ln(P(w_i|\underline{\beta})) = \ln(p(\underline{\beta}|w_i)) + \ln(P(w_i)) - \ln(p(\underline{\beta})) \tag{3.15A}$$

Note that the last term of the discriminant of Equation 3.15A depends only on $\underline{\beta}$ and not on w_i. This term will be present in all discriminants $d_i(\underline{\beta})$, $i = 1,2, \ldots ,M$. Since we are looking for the largest $d_i(\underline{\beta})$, any common term can be ignored. We shall therefore define the discriminant without the last term:

$$d_i(\underline{\beta}) = \ln(p(\underline{\beta}|w_i)) + \ln(P(w_i)) \tag{3.15B}$$

Consider the case where the features are normally distributed. The probability distribution of a signal belonging to the *ith* class, w_i, represented by, $\underline{\beta}$, is

$$p(\underline{\beta}|w_i) = (2\Pi^{-n/2})|\Sigma_i|^{-1/2} \exp\left(-\frac{1}{2}(\underline{\beta} - \mu_i)^T \Sigma_i^{-1}(\underline{\beta} - \mu_i)\right) \tag{3.16}$$

where $\underline{\mu}_i = E\{\underline{\beta}\}$ is the expectation of the *ith* class and Σ_i its n · n convariance matrix:

$$\Sigma_i = E\{(\underline{\beta} - \underline{\mu}_i)(\underline{\beta} - \underline{\mu}_i)^T\} \tag{3.17}$$

The discriminant function (Equation 3.15) for this case is

$$d_i(\underline{\beta}) = -\frac{1}{2}\ln|\Sigma_i| - \frac{1}{2}(\underline{\beta} - \underline{\mu}_i)^T \Sigma_i^{-1}(\underline{\beta} - \underline{\mu}_i) + \ln(P(w_i)) \tag{3.18A}$$

The term $-$ n/2 ln (2Π) was dropped for the same reasons discussed above. Equation 3.18A can be rewritten in the form of a quadratic equation:

$$d_i(\underline{\beta}) = \underline{\beta}^T A_i \underline{\beta} + \underline{b}_i^T \underline{\beta} + c_i \tag{3.18B}$$

where

$$A_i = -\frac{1}{2}\Sigma_i^{-1}$$

$$\underline{b}_i = \Sigma_i^{-1}\underline{\mu}_i$$

and

$$c_i = -\frac{1}{2}\underline{\mu}_i^T \Sigma_i^{-1}\underline{\mu}_i - \frac{1}{2}\ln|\Sigma_i| + \ln(P(w_i))$$

The solution of $d_i(\underline{\beta}) = 0$ yields the decision surfaces in the features space. In the general case, these are hyperquadratic surfaces. In the special case where

$$\Sigma_i = \Sigma \text{ for all } i = 1,2,\ldots,M$$

we are dealing with M classes equally distributed, but with different expectations $\underline{\mu}_i$. In this case the first term of Equation 3.18A can also be ignored as well as the first term of Equation 3.18B. The discriminant function becomes linear, with

$$d_i(\underline{\beta}) = \underline{b}_i^T \underline{\beta} + c_i \tag{3.19A}$$

where

$$\underline{b}_i = \Sigma^{-1}\underline{\mu}_i$$

$$c_i = -\frac{1}{2}\underline{\mu}_i^T \Sigma^{-1}\underline{\mu}_i + \ln(P(w_i))$$

or

$$d_i(\underline{\beta}) = -\frac{1}{2}(\underline{\beta} - \underline{\mu}_i)^T \Sigma^{-1}(\underline{\beta} - \underline{\mu}_i) + \ln(P(w_i)) \tag{3.19B}$$

The first term of the right side Equation 3.19B is the square of the Mahalanobis distance. If, in addition, the *a priori* probabilities $P(w_i)$ are equal they can be ignored, and classification is performed by choosing the class with the minimum Mahalanobis distance (between the signal $\underline{\beta}$ and the class mean $\underline{\mu}_i$). If $P(w_i)$ are not equal the distance is biased in favor of the more probable class.

The simplest case is the case where not only all classes have the same covariance matrices, but also the features are statistically independent. For this case,

$$\Sigma_i = \Sigma = \sigma^2 I \text{ for all i} \tag{3.20}$$

The discriminant becomes

$$d_i(\underline{\beta}) = -\frac{1}{2\sigma^2}(\underline{\beta} - \underline{\mu}_i)^T(\underline{\beta} - \underline{\mu}_i) + \ln(P(w_i))$$

$$= -\frac{1}{2\sigma^2}||\underline{\beta} - \underline{\mu}_i||^2 + \ln(P(w_i)) \tag{3.21A}$$

Here we have the discriminant given in terms of the Euclidean distance. The discriminant can also be written as

$$d_i(\underline{\beta}) = \frac{1}{\sigma^2}\underline{\mu}_i^T\underline{\beta} - \frac{\underline{\mu}_i^T\underline{\mu}_i}{2\sigma^2} + \ln(P(w_i)) \tag{3.21B}$$

where in Equation 3.21B terms which are common to all classes were ignored. Note that the only calculations required in Equation 3.21B are the vector multiplications in the first term. The rest are precalculated and stored as constant in the classifier.

Example 3.1

Assume two classes w_1, w_2 given in a two-dimensional feature space (β_1,β_2) as described in Figure 4. The training set consists of five signals from each class. Assume the features are normally distributed with the same covariance matrix. The discriminant given by Equations 3.19A and 3.19B can be used. The training set and four unknown signals (x) are given below.

Since no information is given on the probabilities of the classes, we shall use $P(w_1) = P(w_2) = 0.5$.

w_1	w_2	x
$\underline{\beta}_{1,1}^T = [-0.5,1]$	$\underline{\beta}_{1,2}^T = [-2,0]$	$\underline{\beta}_{1,x}^T = [-1,0]$
$\underline{\beta}_{2,1}^T = [0,2]$	$\underline{\beta}_{2,2}^T = [-0.5,-0.5]$	$\underline{\beta}_{2,x}^T = [0,0]$
$\underline{\beta}_{3,1}^T = [1,1]$	$\underline{\beta}_{3,2}^T = [-1,-1.5]$	$\underline{\beta}_{3,x}^T = [2,-2]$
$\underline{\beta}_{4,1}^T = [0.5,0.5]$	$\underline{\beta}_{4,2}^T = [0.5,-1]$	$\underline{\beta}_{4,x}^T = [2,0]$
$\underline{\beta}_{5,1}^T = [1,-0.5]$	$\underline{\beta}_{5,2}^T = [0,-2]$	

The expectation of the two classes is estimated by:

$$\hat{\underline{\mu}} = \frac{1}{5}\sum_{j=1}^{5}\underline{\beta}_{j,i}; \quad i = 1,2$$

which yields $\hat{\underline{\mu}}_1^T = [0.4,0.8]$ and $\hat{\underline{\mu}}_2^T = [-0.6,-1]$. Since classes are assumed to possess identical covariance matrix, it is estimated by:

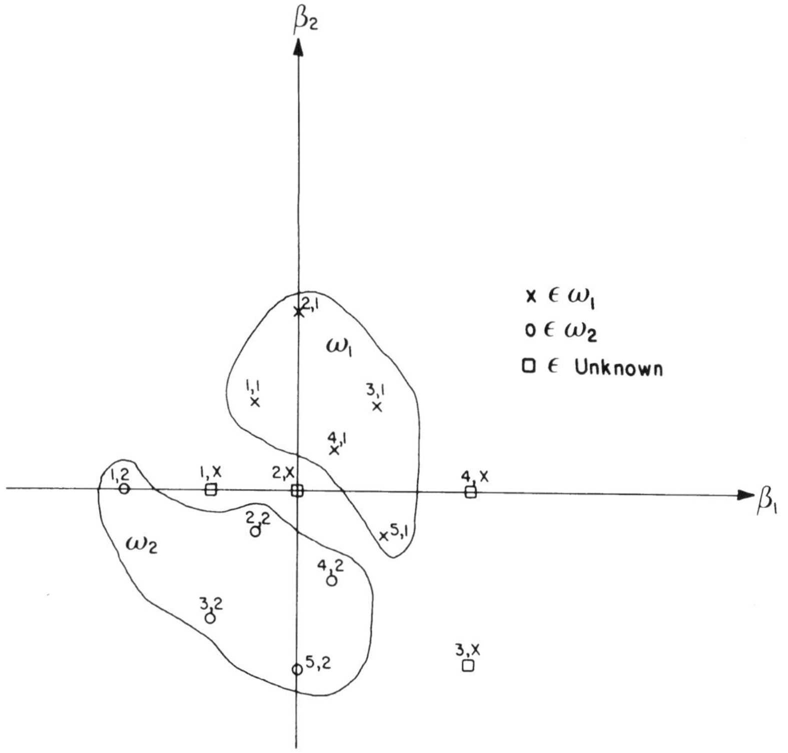

FIGURE 4. Data for Example 1.

$$\hat{\Sigma} = \frac{1}{10}\left[\sum_{j=1}^{5} [(\underline{\beta}_{j,1} - \underline{\hat{\mu}}_1)(\underline{\beta}_{j,1} - \underline{\hat{\mu}}_1)^T + \underline{\beta}_{j,2} - \underline{\hat{\mu}}_2)(\underline{\beta}_{j,2} - \underline{\hat{\mu}}_2)^T]\right]$$

which for the data given is

$$\hat{\Sigma} = \begin{bmatrix} 0.54 & -0.35 \\ -0.35 & 0.68 \end{bmatrix} \quad \hat{\Sigma}^{-1} = \begin{bmatrix} 2.779 & 1.430 \\ 1.430 & 2.207 \end{bmatrix}$$

The discriminant functions $d_1(\underline{\beta})$ and $d_2(\underline{\beta})$ are calculated from Equation 3.19A to give:

$$d_1(\underline{\beta}) = (2.256,\ 2.337)\underline{\beta} - 2.079$$

$$d_2(\underline{\beta}) = (-3.098,\ -3.065)\underline{\beta} - 3.155$$

The discriminant functions for the data set are given below:

$$d_1(\underline{\beta}_{1,1}) = -0.869; \quad d_2(\underline{\beta}_{1,1}) = -4.671$$

$$d_1(\underline{\beta}_{2,1}) = 2.596; \quad d_2(\underline{\beta}_{2,1}) = -9.285$$

$$d_1(\underline{\beta}_{3,1}) = 2.514; \quad d_2(\underline{\beta}_{3,1}) = -9.318$$

$$d_1(\underline{\beta}_{4.1}) = 0.218; \quad d_2(\underline{\beta}_{4.1}) = -6.236$$

$$d_1(\underline{\beta}_{5.1}) = 0.992; \quad d_2(\underline{\beta}_{5.1}) = -4.720$$

$$d_1(\underline{\beta}_{1.2}) = -6.590; \quad d_2(\underline{\beta}_{1.2}) = 3.040$$

$$d_1(\underline{\beta}_{2.2}) = -4.376; \quad d_2(\underline{\beta}_{2.2}) = -0.074$$

$$d_1(\underline{\beta}_{3.2}) = -7.841; \quad d_2(\underline{\beta}_{3.2}) = 4.540$$

$$d_1(\underline{\beta}_{4.2}) = -3.219; \quad d_2(\underline{\beta}_{4.2}) = -1.639$$

$$d_1(\underline{\beta}_{5.2}) = -6.754; \quad d_2(\underline{\beta}_{5.2}) = 2.975$$

The discriminant functions have correctly classified all training data, since $d_1(\underline{\beta}_{j.1}) > d_2(\underline{\beta}_{j.1})$ and $d_1(\underline{\beta}_{j.2}) < d_2(\underline{\beta}_{j.2})$ for all j's. Consider now the four unknown signals $\underline{\beta}_{j.x}$:

$$d_1(\underline{\beta}_{1.x}) = -4.335 \qquad d_2(\underline{\beta}_{1.x}) = -0.057 \rightarrow \underline{\beta}_{1.x} \in w_2$$

$$d_1(\underline{\beta}_{2.x}) = -2.079 \qquad d_2(\underline{\beta}_{2.x}) = 3.155 \rightarrow \underline{\beta}_{2.x} \in w_2$$

$$d_1(\underline{\beta}_{3.x}) = -2.243 \qquad d_2(\underline{\beta}_{3.x}) = -3.220 \rightarrow \underline{\beta}_{3.x} \in w_1$$

$$d_1(\underline{\beta}_{4.x}) = 2.432 \qquad d_2(\underline{\beta}_{4.x}) = -9.285 \rightarrow \underline{\beta}_{4.x} \in w_1$$

Example 3.2

A newborn's cry conveys information concerning different states, needs, and demands of the neonate. It has been demonstrated, by spectrographic studies and by trained listeners, that cry signals originated by birth, hunger, pleasure, and pain have different characteristics. The pain cry of infants with brain damage, hyperbilirubinemia, meningitis, or hypoglycemia has been shown to possess different characteristics. Figure 5 shows typical records of hunger and pain cries from newborn infants. The hunger cry was recorded from a healthy infant awaiting feeding, 4 hr after the previous feeding. The pain cry was recorded during hip joint examination (Ortolani test). Bayes classification was used for automatic classification of the cry signals. An AR model (see Chapter 7, Volume I) of order 20 was used to model the signals. Figure 6 shows the pole map and LPC estimated spectrum of the two cries. The first ten LPCs were chosen as the feature vector:

$$\underline{\beta}^T = [\hat{a}_1, \hat{a}_2, \ldots, \hat{a}_{10}] \tag{3.22}$$

Hunger and pain cry records from five infants were used as a training set. The mean feature vectors for the hunger cry $\underline{\hat{\mu}}_H$, and for the pain cry, $\underline{\hat{\mu}}_P$, were estimated by:

$$\underline{\hat{\mu}}_i = \frac{1}{N_i} \sum_{j=1}^{N_i} \underline{\beta}_i^j$$

$$i = H,P \tag{3.23}$$

where $\underline{\beta}_i^j$ is the *jth* training vector of the $i \in (H,P)$ class. The covariance matrices Σ_H and Σ_p were estimated by:

$$\hat{\Sigma}_i = \frac{1}{N_i}\sum_{j=1}^{N_i}(\underline{\beta}_i^j - \underline{\hat{\mu}}_i)(\underline{\beta}_i^j - \underline{\hat{\mu}}_i)^T$$

$$i = H,P \tag{3.24}$$

The Bayes rule (Equation (3.5)) in its log form with the probability of Equation 3.16 for this case becomes:

$$\frac{1}{2}(\underline{\beta} - \underline{\hat{\mu}}_H)^T \hat{\Sigma}_H^{-1}(\underline{\beta} - \underline{\hat{\mu}}_H) - \frac{1}{2}(\underline{\beta} - \underline{\mu}_P)^T \Sigma_P^{-1}(\underline{\beta} - \underline{\hat{\mu}}_P)$$

$$+ \frac{1}{2}\ln\frac{|\hat{\Sigma}_H|}{|\hat{\Sigma}_P|} \lessgtr \ln\left(\frac{P(w_H)}{P(w_P)}\right) \rightarrow \underline{\beta} \in \begin{cases} w_H \\ w_P \end{cases} \tag{3.25}$$

where w_H and w_P denote the hunger and pain classes, respectively. Define the quadratic distance, $D_{j,\underline{\beta}}$ (Mahalanobis distance):

$$D_{j,\underline{\beta}} = (\underline{\beta} - \underline{\hat{\mu}}_j)^T \Sigma_j^{-1}(\underline{\beta} - \underline{\hat{\mu}}_j)$$

$$j = H,P \tag{3.26}$$

and

$$D = D_{H,\underline{\beta}} - D_{P,\underline{\beta}} \tag{3.27}$$

and

$$THR = 2\ln\left(\frac{p(w_H)}{p(w_P)}\right) - \ln\left(\frac{|\Sigma_H|}{|\Sigma_P|}\right) \tag{3.28}$$

Then the decision rule (Equation 3.25) can be rewritten:

$$D \lessgtr THR \rightarrow \underline{\beta} \in \begin{cases} w_H \\ w_P \end{cases} \tag{3.29}$$

Equation 3.29 is the quadratic Bayes test for minimum error. This classifier does not allow rejects. Each data record is forced to be classified even if it is a "bad" record in the sense that it includes artifacts or does not belong to each one of the two classes. Consider now the case where we introduce two threshold levels, R_1 and R_2, such that:

$$D = D_{H,\underline{\beta}} - D_{P,\underline{\beta}} \begin{cases} \leq R_1 & \rightarrow \underline{\beta} \in w_H \\ \geq R_2 & \rightarrow \underline{\beta} \in w_P \\ \text{otherwise} & \rightarrow \underline{\beta} \in \text{Reject} \end{cases} \tag{3.30}$$

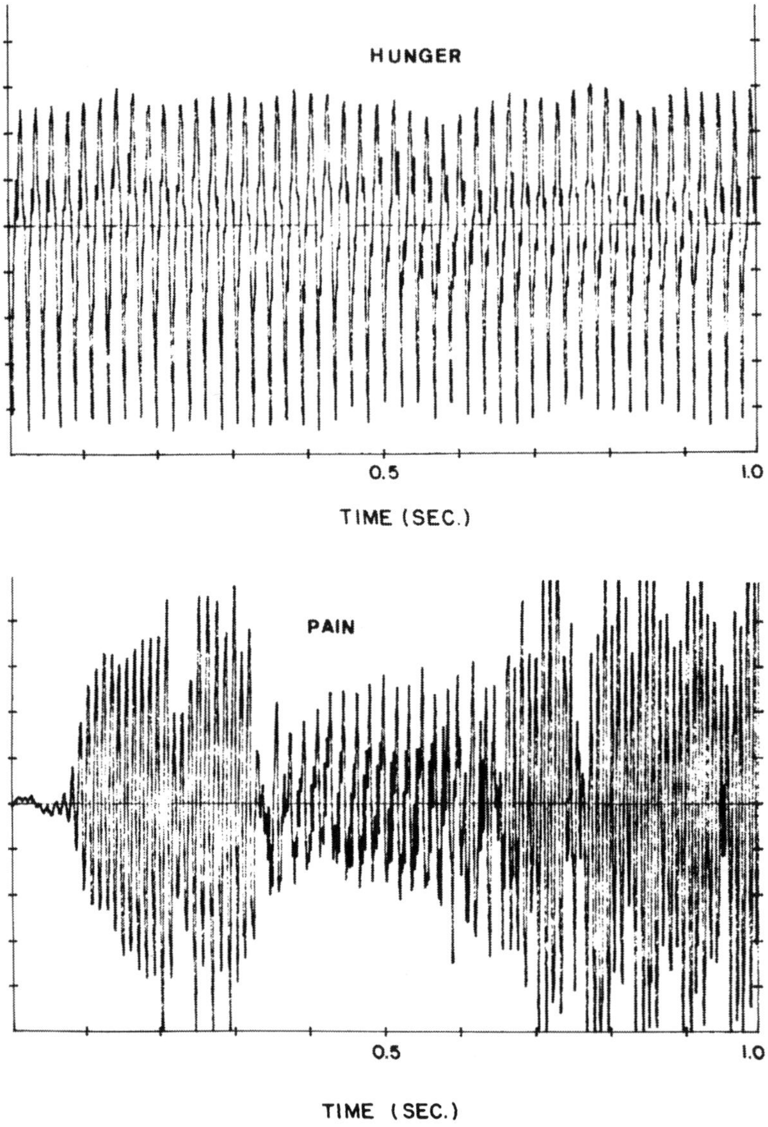

A

FIGURE 5. Typical hunger and pain cry. (A) Time Records; (B) power spectral density estimated by FFT.

With no rejections, two types of classification errors were present: (1) errors due to the classification of pain cry as hunger cry (denote the probability of this error by $\tilde{\epsilon}_{H|p}$):

$$\tilde{\epsilon}_{H|P} = \int_{-\infty}^{THR} p(D|\underline{\beta} \in w_p)dD \tag{3.31A}$$

and (2) errors due to the classification of hunger cry as pain cry (with the probability, $\tilde{\epsilon}_{p|H}$):

$$\tilde{\epsilon}_{P|H} = \int_{THR}^{\infty} p(D|\underline{\beta} \in w_H)dD \tag{3.31B}$$

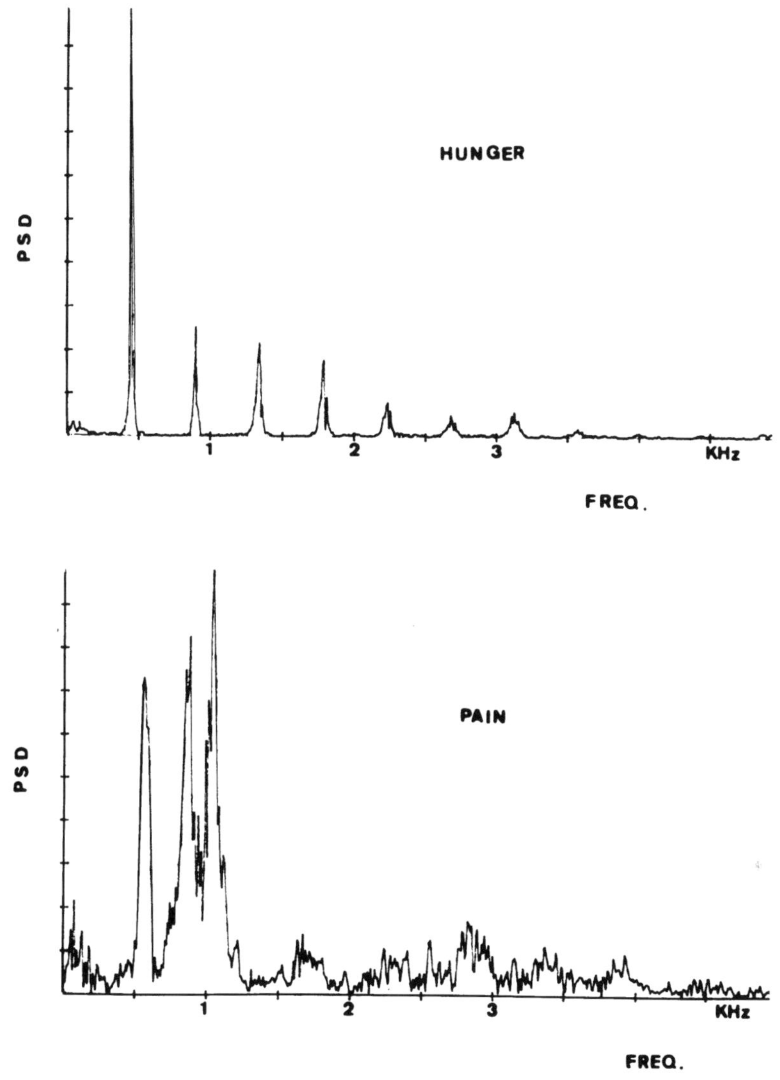

FIGURE 5B.

With rejection, using decision rule (Equation 3.30), the errors are

$$\epsilon_{H|P} = \int_{-\infty}^{R_1} p(D|\underline{\beta} \in w_p)dD \leq \bar{\epsilon}_{H|P} \tag{3.31C}$$

and

$$\epsilon_{P|H} = \int_{R_2}^{\infty} p(D|\underline{\beta} \in w_H)dD \leq \bar{\epsilon}_{P|H} \tag{3.31D}$$

However, rejections will be present. Some of the rejections were correctly classified by Equation 3.29. The probability of a hunger cry record, correctly classified by Equation 3.29 and rejected by Equation 3.30, is

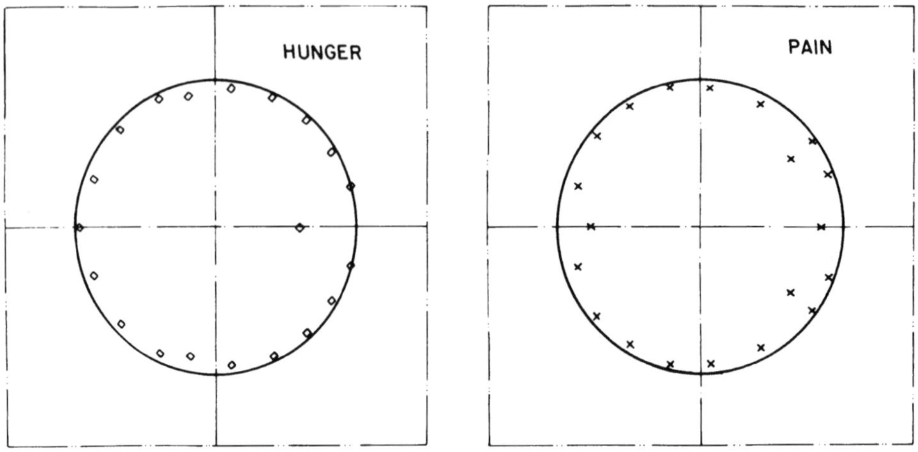

FIGURE 6. Poles maps of hunger and pain cries.

$$\epsilon^R_{H|H} = \int_{R_1}^{THR} p(D|\underline{\beta} \in w_H)dD \tag{3.31E}$$

and similarly for pain cry:

$$\epsilon^R_{p|p} = \int_{THR}^{R_2} p(D|\underline{\beta} \in w_p)dD \tag{3.31F}$$

The probability densities, $p(D|\underline{\beta})$, are estimated from the training data. We shall choose the rejection threshold R_1 such as to minimize a linear combination of the error probabilities $\epsilon^R_{H|H}$ and $\epsilon_{H|P}$. Similar considerations will dictate the decision for R_2. Hence,

$$R_1 \rightarrow \underset{R_1}{\mathrm{Min}}(\omega_1\epsilon^R_{H|H} + \omega_2\epsilon_{H|p})$$

$$R_2 \rightarrow \underset{R_2}{\mathrm{Min}}(\omega_3\epsilon^R_{p|p} + \omega_3\epsilon_{p|H}) \tag{3.32}$$

where ω_i, i = 1,2,3,4, are weights, determined by the relative importance of each one of the errors. Figure 7 depicts the training and classification system. Automatic classification of infant's cry was performed by the classifier (Equation 3.30), with error rate of less than 5%.[38] The quadratic classifier (Equation 3.29), with no rejection, was also applied to the classification of single evoked potentials[39] with similar results.

C. k-Nearest Neighbor (k-NN) Classification

In most practical cases, the statistics of the features are unknown. The classification rules discussed before can be applied therefore only after the probabilities have been estimated from the given data.

The probability density of the features vector $\underline{\beta}$ is to be estimated. Consider a given region R in the features space; the probability P that the vector $\underline{\beta}$ will fall in this region is

$$P = \int_R p(\underline{\beta})d\underline{\beta} \approx p(\underline{\beta})V \tag{3.33}$$

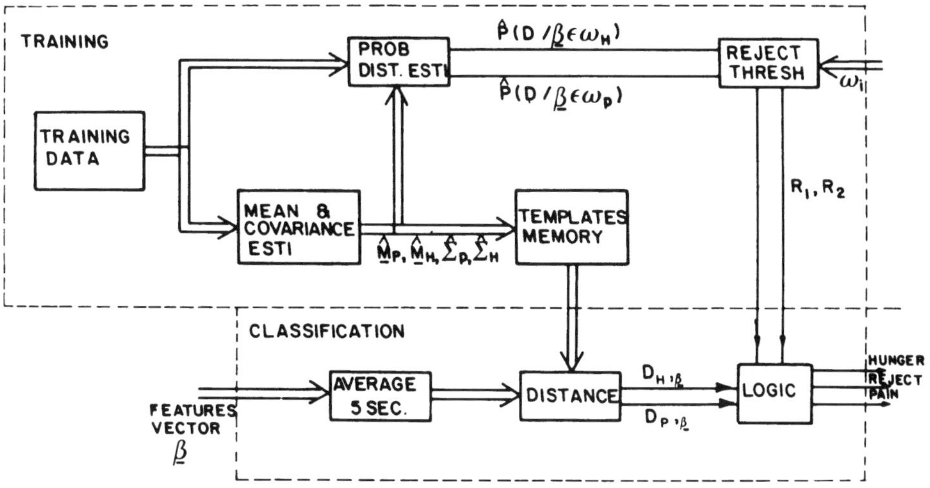

FIGURE 7. Training and classification system for classifier with rejection.

where V is the volume of R.

The right side of the equation holds if R is very small. Suppose that we have drawn m samples of the feature vector, k of which fell in the region R. The probability P can be estimated by k/m; hence,

$$\hat{p}(\underline{\beta}) = \frac{k/m}{V} \tag{3.34}$$

In Equation 3.34 one must decide the size of V to be used. Clearly V cannot be allowed to grow since this will cause the estimation to be "smoothed". On the other hand, V cannot be too small since then the variance of the estimate will increase. One method for choosing the volume V is to determine k as some function of m such that, for example, $k = k_m = k \cdot \sqrt{m}$. The volume chosen, denoted by V_m, is determined by increasing V until it contains k_m neighbors of $\underline{\beta}$. This is known as the k_m-nearest neighbor estimation. Note that the volume chosen for the estimation of the probability functions becomes a function of the data. If the features are densely distributed around $\underline{\beta}$, the volume containing k_m neighbors will be small.

Assume that the training set consists of total $N = \sum_{i=1}^{M} N_i$ samples of feature vectors, where N_i is the number of samples belonging to the class w_i. An unknown signal, represented by the vector, $\underline{\beta}$, is to be classified. We find the volume V_m around $\underline{\beta}$ that includes k_m training samples. Out of the k_m samples, k_i samples belong to the class w_i. An estimate for the conditional probability becomes:

$$\hat{p}(\underline{\beta}|w_i) = \frac{k_i/N_i}{V}$$

and

$$\hat{P}(w_i) = \frac{N_i}{N} \tag{3.35}$$

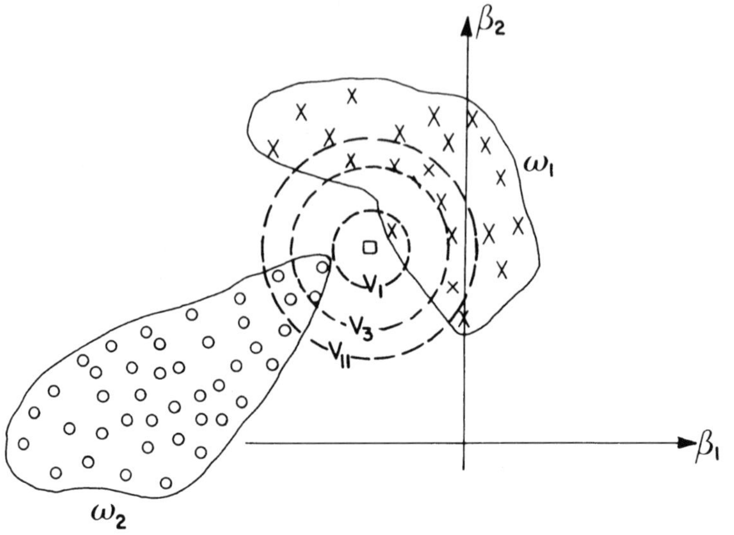

FIGURE 8. k-Nearest neighbor (k-NN) classification.

The Bayes minimum error rate rule (Equation 3.14) can be employed using the Bayes relation (Equation 3.1) and the estimates of Equation 3.35. Namely, classify $\underline{\beta}$ into w_i if:

$$\frac{N_i}{N}\,\hat{p}(\underline{\beta}|w_i) > \frac{N_j}{N}\,\hat{p}(\underline{\beta}|w_j) \text{ for all j} \tag{3.36A}$$

or by substituting Equation 3.35 if:

$$k_i > k_j \text{ for all j} \tag{3.36B}$$

or

$$k_i = \text{Max}\{k_1,k_2,\ldots,k_M\} \rightarrow \underline{\beta} \in w_i \tag{3.36C}$$

For the case of two classes, the k-nearest neighbor rule becomes:

$$k_1 \gtrless k_2 \rightarrow \underline{\beta} \in \begin{cases} w_1 \\ w_2 \end{cases} \tag{3.36D}$$

This rule does not require the knowledge of the features statistics and is relatively simple to use. It does, however, require the storage of all training samples. Figure 8 demonstrates the k-nearest neighbor classification rule. It shows a two-dimensional case with two classes, $w_1(x)$ and $w_2(o)$. The unknown signal is represented by the features vector $\underline{\beta}$ (□). Choosing $k = 1$, we see that the classification is $\underline{\beta} \in w_1$ since in V_1 the sample encountered belongs to w_1. For $k = 3$, the classification is $\underline{\beta} \in w_2$, for $k = 11$, $\underline{\beta} \in w_1$.

It can be shown[3] that the k-nearest neighbor classification error, ϵ_{NN}, is related to the Bayes classification error, ϵ_B, by the inequality:

$$\epsilon_{NN} \leqslant 2\epsilon_B(1 - \epsilon_B) \tag{3.37}$$

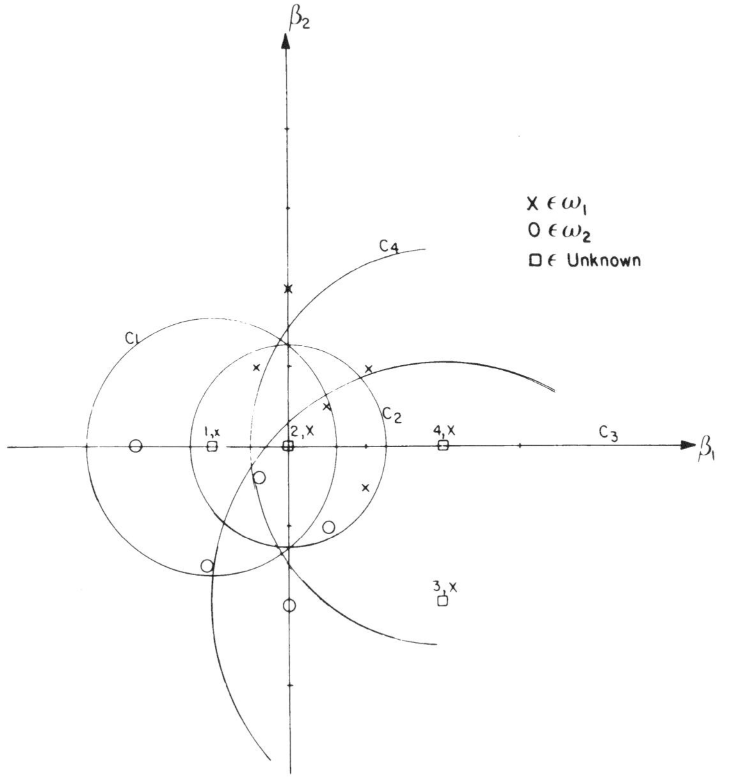

FIGURE 9. k-NN classification of Example 1.

Since $\epsilon_B \ll 1$, we note that the upper limit for the k-NN error is twice that of the Bayes error. The k-NN classification method has been applied, for example, to the problem of automatic classification of EEG signals.[15]

Example 3.3

Consider the data of Example 1 shown in Figure 4. We want to classify the unknown signals $\underline{\beta}_{j,x}$, $j = 1,2,3,4$, by means of the k-NN method with $k = 5$. Refer to Figure 9 where the circles c_j determine the volumes including five nearest neighbors. The classification is

$$\underline{\beta}_{1,x} \in w_2,\ \underline{\beta}_{2,x} \in w_1,\ \underline{\beta}_{3,x} \in w_2,\ \underline{\beta}_{4,x} \in w_1$$

note that $\underline{\beta}_{2,x}$ was classified as belonging to w_1 while it was classified to w_2 using the discriminant in Example 3.1. $\underline{\beta}_{3,x}$ was classified here as belonging to w_2 rather to w_1.

III. LINEAR DISCRIMINANT FUNCTIONS

A. Introduction

Linear discriminant functions have several advantages. They are relatively easy to compute and simple to operate. Refer again to Figure 2. A linear function in this plane (β_1,β_2) can be shown to discriminate between the classes w_1 and w_2. We can write the linear disciminant function:

$$d(\underline{\beta}) = \rho_1\beta_1 + \rho_2\beta_2 + \rho_3 = 0$$

where ρ_i, i = 1,2,3, are weights which are the parameters of the linear disciminant function. In general consider the n*th* dimensional features space — the linear discriminant function $d(\underline{\beta}) = 0$ describes a hyperplane in the space:

$$d(\underline{\beta}) = \sum_{i=1}^{n} \rho_i\beta_i + \rho_{n+1} = \underline{\rho}^T\underline{\beta} + \rho_{n+1} = 0 \tag{3.38}$$

where the weighting vector $\underline{\rho}^T = [\rho_1,\rho_2, \ldots ,\rho_n]$ and the features vector $\underline{\beta}^T = [\beta_1,\beta_2, \ldots ,\beta_n]$. Define the augmented vectors $\underline{\rho}_A^T = [\rho_1,\rho_2, \ldots ,\rho_n,\rho_{n+1}]$ and $\underline{\beta}_A^T = [\beta_1,\beta_2, \ldots ,\beta_n,1]$; then Equation 3.38 can be rewritten:

$$d(\underline{\beta}) = \underline{\rho}_A^T\underline{\beta}_A = 0 \tag{3.39}$$

The classification decision in the case of two classes is done by:

$$d(\underline{\beta}) = \underline{\rho}_A^T\underline{\beta}_A \gtrless 0 \rightarrow \underline{\beta} \in \begin{cases} w_1 \\ w_2 \end{cases} \tag{3.40}$$

which means that if for a given feature vector $\underline{\beta}$ the discriminant $d(\underline{\beta})$ is larger than zero, the signal is classified into w_1. If $d(\underline{\beta}) = 0$, the classification is undetermined. In the case of more than two classes, say M classes, w_i, i = 1,2, . . . ,M, we have M weighting vectors $\underline{\rho}_i$ and M discriminant functions $d_i(\underline{\beta})$. The space is thus divided into (at least) M regions, Ω_i, i = 1,2, . . . ,M. Each region is defined by

$$\Omega_i\text{: } d_i(\underline{\beta}) > 0 \text{ and } d_j(\underline{\beta}) < 0$$

$$j \neq i \tag{3.41}$$

An example with M = 3 is shown in Figure 10. The shaded areas define the classification regions, Ω_i.

A geometrical meaning can be attached to the discriminant function $d(\underline{\beta})$. Consider the example in Figure 11, where the discriminant function $d(\underline{\beta})$ is shown as a line in the two-dimensional feature space. Consider the vector $\underline{n}$, a unit length vector normal to the hyperplane $d(\underline{\beta})$. It is easy to show that

$$\underline{n} = \frac{\underline{\rho}}{||\underline{\rho}||} \tag{3.42}$$

Consider a signal represented in the feature space by the vector $\underline{\beta}_0$. We can write the vector $\underline{\beta}_0$ as:

$$\underline{\beta}_0 = \underline{\beta}_d + r\underline{n} = \underline{\beta}_d + r\frac{\underline{\rho}}{||\underline{\rho}||} \tag{3.43}$$

where $\underline{\beta}_d$ is the normal projection of $\underline{\beta}_0$ on the hyperplane and r is the distance between the point $\underline{\beta}_0$ and the hyperplane. Let us find the value of the discriminant function at $\underline{\beta}_0$, from Equation 3.38:

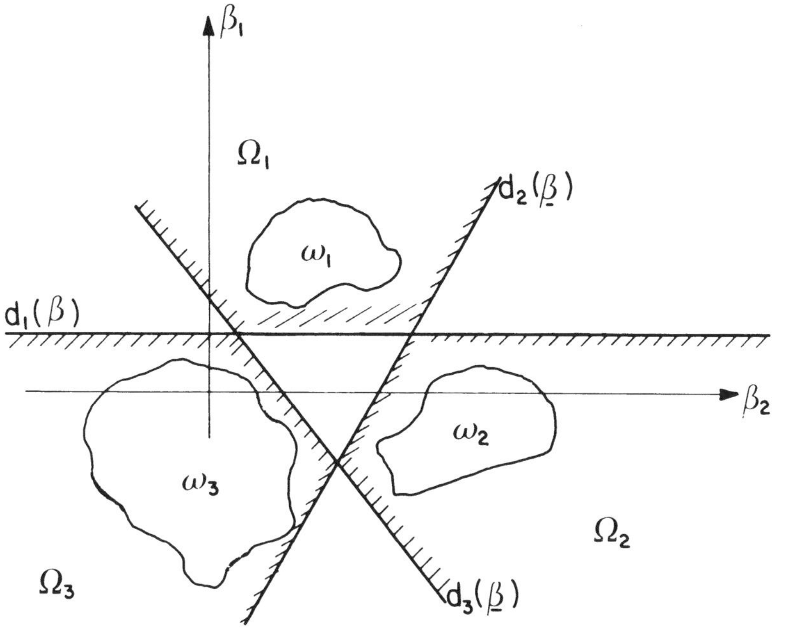

FIGURE 10. Linear discriminant functions.

$$d(\underline{\beta}_0) = \underline{\rho}^T\underline{\beta}_0 + \rho_{n+1} = \underline{\rho}^T\underline{\beta}_d + \underline{\rho}^T r \frac{\underline{\rho}}{||\underline{\rho}||} + \rho_{n+1} \tag{3.44}$$

Since $\underline{\beta}_d$ is on the hyperplane, we have

$$d(\underline{\beta}_d) = 0 = \underline{\rho}^T\underline{\beta}_d + \rho_{n+1}$$

Hence, we get

$$d(\underline{\beta}_0) = \underline{\rho}^T r \frac{\underline{\rho}}{||\underline{\rho}||} = r\,||\underline{\rho}|| \tag{3.45A}$$

or

$$r = \frac{d(\underline{\beta}_0)}{||\underline{\rho}||} \tag{3.45B}$$

Recall that r is the distance between the point $\underline{\beta}_0$ and the hyperplane. The discriminant function is thus related to that distance.

B. Generalized Linear Discriminant Functions

In Equation 3.39, the discriminant function was defined as a linear combination of the features β_i. This can be generalized by defining it as a linear combination of functions of $\underline{\beta}$. Consider the new definition:

$$\begin{aligned} d(\underline{\beta}) &= \rho_1 f_1(\underline{\beta}) + \rho_2 f_2(\underline{\beta}) + \ldots + \rho_n f_n(\underline{\beta}) + \rho_{n+1} \\ &= \sum_{i=1}^{n+1} \rho_i f_i(\underline{\beta}) \end{aligned} \tag{3.46}$$

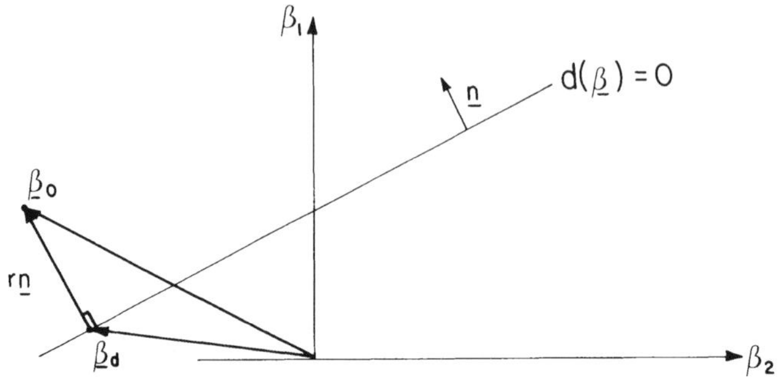

FIGURE 11. Geometry of the discriminant hyperplane.

where $f_{n+1}(\underline{\beta}) = 1$. Define the vector $\underline{\beta}_f$:

$$\underline{\beta}_f^T = [f_1(\underline{\beta}), f_2(\underline{\beta}), \ldots, f_n(\underline{\beta}), 1] \tag{3.47}$$

Then Equation 3.46 can be expressed by

$$d(\underline{\beta}) = \underline{\rho}_A^T \underline{\beta}_f \tag{3.48}$$

Note that while the functions $f_i(\underline{\beta})$ can be nonlinear, Equation 3.48 is linear with respect to $\underline{\beta}_f$.

Consider, for example, the case where the functions $f_i(\underline{\beta})$ are quadratic functions of the form:

$$f_i(\underline{\beta}) = \beta_k \beta_\ell$$

$$k, \ell = 1, 2, \ldots, n$$

In this case and for the two-dimensional feature space, the discriminant is

$$d(\underline{\beta}) = \rho_{11}\beta_1^2 + \rho_{12}\beta_1\beta_2 + \rho_{22}\beta_2^2 + \rho_1\beta_1 + \rho_2\beta_2 + \rho_3$$

For the general n dimensional space,

$$d(\underline{\beta}) = \sum_{j=1}^{n} \rho_{jj}\beta_j^2 + \sum_{j=1}^{n-1} \sum_{k=j+1}^{n} \rho_{jk}\beta_j\beta_k + \sum_{j=1}^{n} \rho_j\beta_j + \rho_{n+1} \tag{3.49}$$

which can be written in a matrix form as:

$$d(\underline{\beta}) = \underline{\beta}^T A \underline{\beta} + \underline{\beta}^T \underline{b} + c \tag{3.50}$$

The matrix A vector $\underline{b}$ and c determine the hyperplane. Equation 3.50 uses weights and functions for the discriminant. For a given problem, a set of functions has to be determined and the proper weights found.

C. Minimum Squared Error Method

The decision surfaces $d(\underline{\beta}) = 0$ have to be defined such that the classification error be minimum. This calls for some optimal method to determine the weighting vector, $\underline{\rho}$.

Consider the two-class problem. Assume $\underline{\beta}_i \in w_1$; then it is correctly classified if:

$$d(\underline{\beta}_i) = \underline{\rho}_A^T\underline{\beta}_i > 0 \quad (3.51A)$$

and $\underline{\beta}_i \in w_2$ if:

$$d(\underline{\beta}_i) = \underline{\rho}_A^T\underline{\beta}_i < 0 \quad (3.51B)$$

We can replace all $\underline{\beta}$'s belonging to w_2 with $(-\underline{\beta})$. The classification rule for w_2 becomes similar to Equation 3.51A.

Several methods are used to find the optimal weighting function for Equation 3.51A. Among them are gradient descent procedures, the perception criterion function, and various relaxation procedures. We shall present here the method of minimum squared error. Instead of solving the linear inequalities (Equation 3.51A), we shall look for the solution of the set of linear equations:

$$\underline{\rho}_A^T\underline{\beta}_i = b_i \qquad i = 1,2,\ldots,N \quad (3.52)$$

where b_i, $i = 1,2, \ldots ,N$ are some arbitrarily chosen positive constants known as the "margins". The set of N equations (Equation 3.52) is solved to determine the weighting vector $\underline{\rho}_A$, in such a way that the N known samples will be classified with minimum error.

Define the $Nx(n + 1)$ matrix F:

$$F^T = [\underline{\beta}_1 \vdots \underline{\beta}_2 \vdots, \ldots, \vdots \underline{\beta}_N] \quad (3.53)$$

and define the constants vector $\underline{b}^T = [b_1,b_2, \ldots ,b_N]$; then Equations 3.52 becomes:

$$F\underline{\rho}_A = \underline{b} \quad (3.54)$$

Note that the matrix F is not square; hence Equation 3.54 cannot be solved directly. The pseudo inverse of F must be used. Define the error vector:

$$\underline{e} = F\underline{\rho}_A - \underline{b} \quad (3.55)$$

and find $\underline{\rho}_A$ that minimizes the sum squares of $\underline{e}$:

$$\underset{\underline{\rho}_A}{\text{Min}}\ \underline{e}^T\underline{e} = \underset{\underline{\rho}_A}{\text{Min}}(F\underline{\rho}_A - \underline{b})^T(F\underline{\rho}_A - \underline{b}) = \underset{\underline{\rho}_A}{\text{Min}}(\underline{\rho}_A^T F^T F \underline{\rho}_A - 2\underline{\rho}_A^T F^T \underline{b}) \quad (3.56)$$

The solution of Equation 3.56 yields the least squares estimate:

$$\underline{\rho}_A = (F^TF)^{-1}\, F^T\underline{b} \quad (3.57A)$$

Note that the $(n + 1)\cdot(n + 1)$ matrix F^TF is square. It may, however, be ill conditioned; in such cases the ridge regression method should be used such that:

$$\underline{\rho}_A = \lim_{\epsilon \to 0}(F^TF + \epsilon I)^{-1}\, F^T\underline{b} \quad (3.57B)$$

The solution (Equation 3.57) depends on the margin vector $\underline{b}$. Different choices of $\underline{b}$ will lead to different decision surfaces.

D. Minimum Distance Classifiers

In cases where the various classes are well clustered and separated from one another, classification can be performed based on the proximity of the unknown vector to the various clusters. The unknown vector is classified into the cluster to which it is closest. For such a classification scheme we need to define a prototype to represent each one of the clusters (classes) and a scalar measure of proximity. We call the measure of proximity a "distance" and the classification scheme "minimum distance classification".

Consider the case with M classes each represented by a template, or a prototype, $\underline{m}_i$, i = 1,2, . . . ,M. The template can be, for example, some weighted average of the class training set. The distance measure can be defined in various ways. Consider first the Euclidean distance, D_i, between the unknown signal to be classified $\underline{\beta}$ and the *ith* template $\underline{m}_i$:

$$D_i^2 = \|\underline{\beta} - \underline{m}_i\|^2 = (\underline{\beta} - \underline{m}_i)^T(\underline{\beta} - \underline{m}_i) \tag{3.58}$$

We can choose D_i^2 to be the distance measure, since all distances are positive. To classify, $\underline{\beta}$, we calculate Equation 3.58 for all i and choose the class j that yields the smallest distance:

$$D_j^2 = \underset{i}{\mathrm{Min}}(D_i^2) \rightarrow \underline{\beta} \in w_j \tag{3.59}$$

A closer look[4] at Equation 3.58 will show the relation between the minimum distance classifier and linear discrimination:

$$D_i^2 = (\underline{\beta} - \underline{m}_i)^T(\underline{\beta} - \underline{m}_i) = \underline{\beta}^T\underline{\beta} - 2\left(\underline{\beta}^T\underline{m}_i - \frac{1}{2}\underline{m}_i^T\underline{m}_i\right) \tag{3.60}$$

The first term on the right side is independent of i and thus can be ignored in the minimization process. Minimization of D_i^2 is the same as maximization of the second term of the right side. Hence, we can define an equivalent decision function:

$$d_i(\underline{\beta}) = \underline{\beta}^T\underline{m}_i - \frac{1}{2}\underline{m}_i^T\underline{m}_i \qquad i = 1,2,\ldots,M \tag{3.61}$$

Define a weighting vector $\underline{\rho}_i$ such that:

$$\underline{\rho}_i^T = \left[\underline{m}_i^T \vdots \left(-\frac{1}{2}\underline{m}_i^T\underline{m}_i\right)\right] \tag{3.62A}$$

and the augmented feature vector, $\underline{\beta}_A$:

$$\underline{\beta}_A^T = [\underline{\beta}^T \vdots 1] \tag{3.62B}$$

Then the decision function (Equation 3.61) becomes:

$$d_i(\underline{\beta}) = \underline{\rho}_i^T\underline{\beta}_A \qquad i = 1,2,\ldots,M \tag{3.63}$$

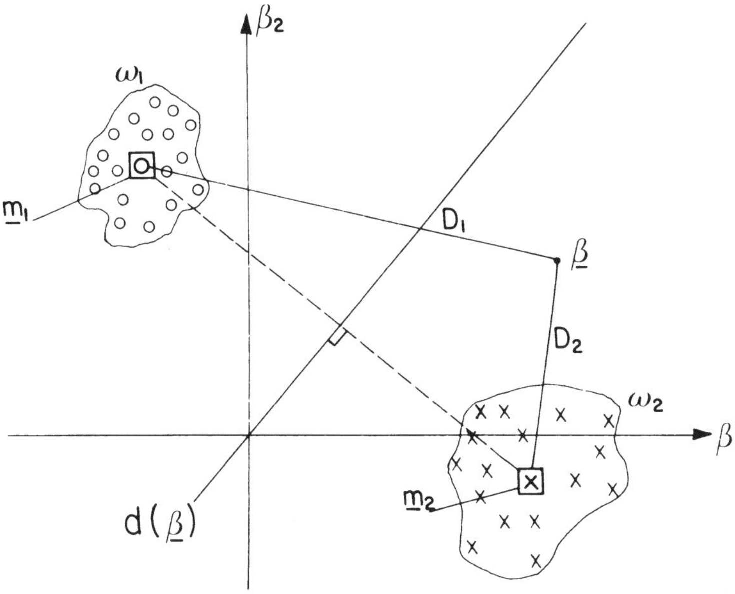

FIGURE 12. Minimum distance classifier.

which is a linear discriminant function discussed in Section III. Figure 12 shows a simple case of two classes in a two-dimensional features space. It can be shown that in this case the decision surface, $d(\underline{\beta}) = 0$, is a hyperplane normal to the line segment joining the two templates, located at an equal distance from them.

Note that the Euclidean distance of Equation 3.58 gives equal importance to each one of the elements of the features space. If, however, we have some *a priori* knowledge on the statistics of the classes, we may want to place weights on the features. If, for example, it is *a priori* known that some of the features have large variances, we may want to consider them "less reliable" when defining the proximity measure. This leads intuitively to a distance measure where the weights are inversely proportional to the features covariance matrix, namely,

$$D_i^2 = (\underline{\beta} - \underline{m}_i)^T \Sigma_i^{-1} (\underline{\beta} - \underline{m}_i) \tag{3.64}$$

which is the Mahalanobis distance discussed previously (Equation 3.19). Note that if Equation 3.64 is used, the problem is a quadratic one.

Example 3.4

Consider the minimum distance classification for the data of Example 1 (Figure 4). The Euclidean distance (Equation 3.58) for the four unknown signals is calculated below:

	D_1^2	D_2^2
$\underline{\beta}_{1,x} \rightarrow$	2.6	$1.16 \rightarrow \underline{\beta}_{1,x} \in w_2$
$\underline{\beta}_{2,x} \rightarrow$	0.8	$1.36 \rightarrow \underline{\beta}_{2,x} \in w_1$
$\underline{\beta}_{3,x} \rightarrow$	10.4	$7.76 \rightarrow \underline{\beta}_{3,x} \in w_2$
$\underline{\beta}_{4,x} \rightarrow$	3.2	$7.76 \rightarrow \underline{\beta}_{4,x} \in w_1$

The signals were classified as in the 5-NN (Example 3.3).

E. Entropy Criteria Methods

1. Introduction

The amount of information content is often measured by the entropy, which is a statistical measure of uncertainty.[4] Consider the probability density function, $p(\beta)$, of the features population.

The entropy, $H(\underline{\beta})$, is given by (see also Chapter 8, Volume I, Section III)

$$H(\underline{\beta}) = -E\{\ln(p(\underline{\beta}))\} = -\int p(\underline{\beta})\ln(p(\underline{\beta}))d\underline{\beta} \tag{3.65}$$

Consider M classes, w_i, $i = 1,2, \ldots, M$. We shall make the restrictive assumption that the signals of all M classes are normally distributed. The signals belonging the *ith* class have the expectation $\underline{\mu}_i$ and covariance matrix, Σ_β. We assume therefore that all classes have the same covariance matrix. This may be the case, for example, where the classes are deterministic vectors $\underline{\mu}_i$, $i = 1,2, \ldots, M$, and the measured signals are noisy signals from zero mean normally distributed noise process common to all measurements.

The conditional probabilities $p(\underline{\beta}|w_i)$, $i = 1,2, \ldots, M$, are given by Equation 3.16. The entropy of w_i is

$$H_i(\underline{\beta}) = -\int p(\underline{\beta}|w_i)\ln(p(\underline{\beta}|w_i))d\underline{\beta} \tag{3.66}$$

where the integration is performed over the features space. We would like now to find a linear transformation:

$$\underline{y} = T^T\underline{\beta} \tag{3.67}$$

that transfers the n dimensional feature vector, $\underline{\beta}$, into a reduced dimensional vector $\underline{y}$. The transformation matrix T is thus of dimension $n \times d$. This is a procedure commonly taken when the goal is signal compression.

2. Minimization of Entropy

Here we shall look for the transformation that not only reduces the dimensionality of the problem, but mainly preserves or even enhances the discrimination properties between classes. The entropy is a measure of uncertainty since features which reduce the uncertainty of a given situation are considered more informative. Tou[4,46] has suggested that minimization of the entropy is equivalent to minimizing the dispersion of the various classes, while not effecting much the interclass dispersion. It is thus reasonable to expect that the minimization will have clustering properties.

The $d \times d$ covariance matrix, Σ_y, of the reduced vector $\underline{y}$, whose expectation is $\underline{\tilde{\mu}}$, is given by

$$\Sigma_y = E\{(\underline{y} - \underline{\tilde{\mu}})(\underline{y} - \underline{\tilde{\mu}})^T\} = T^T \Sigma_\beta T \tag{3.68}$$

Since the new vector, $\underline{y}$, is the result of a linear transformation of a gaussian process, it is also gaussian. The conditional probability density in the reduced space is thus:

$$p(\underline{y}|w_i) = (2\Pi)^{d/2}|\Sigma_y|^{-1/2} \exp\left(-\frac{1}{2}(\underline{y} - \underline{\tilde{\mu}})^T \Sigma_y^{-1}(\underline{y} - \underline{\tilde{\mu}})\right) \tag{3.69}$$

and the entropy in the reduced feature space:

$$H_{(y)} = -\int p(\underline{y}|w_i)\ln(p(\underline{y}|w_i)d\underline{y} = \frac{d}{2} + \frac{1}{2}\ln|\Sigma_y| + \frac{d}{2}\ln(2\Pi) \tag{3.70A}$$

The determinant of the covariance matrix in Equation 3.70A is equivalent to the product of its eigenvalues; hence,

$$H_{(y)} = \frac{1}{2}\sum_{i=1}^{d} \ln \lambda_i + \frac{d}{2}(1 + \ln(2\Pi)) \tag{3.70B}$$

where λ_i are the eigenvalues of the covariance matrix Σ_y. The minimization[46] of Equation 3.70b yields the optimal transformation T, which turns out to be a matrix whose d columns are the d eigenvectors of Σ_β associated with the *smallest* eigenvalues $\lambda_{n-d+1}, \lambda_{n-d+2}, \ldots, \lambda_n$ of Σ_β. Hence, the entropy of the reduced vector is

$$H_{(y)} = \frac{1}{2}\sum_{i=n-d+1}^{n} \ln \lambda_i + \frac{d}{2}(1 + \ln(2\Pi)) \tag{3.70C}$$

In Equation 3.70C we have assumed that the eigenvalues of Σ_β had been arranged in decreasing order.

Example 3.5

Consider again the two classes of Example 3.1 depicted in Figure 4. It is required to apply the minimum entropy criterion to reduce dimensionality while preserving clustering. The estimation of the covariance matrix was calculated in Example 3.1. The eigenvalues are given by

$$\det(\hat{\Sigma} - \lambda I) = |\hat{\Sigma} - \lambda| = 0 = \begin{vmatrix} 0.54 - \lambda & -0.35 \\ -0.35 & 0.68 - \lambda \end{vmatrix}$$

The solution yields the two eigenvalues, $\lambda_1 = 0.96693$ and $\lambda_2 = 0.25307$. The corresponding eigenvectors $\underline{u}_i$ are given by:

$$\hat{\Sigma}\underline{u}_i = \lambda_i\underline{u}_i$$

$$i = 1,2$$

The solution yields

$$\underline{u}_1^T = [0.63399, \ -0.77334]; \quad \underline{u}_2^T = [0.77334, \ 0.63399]$$

The required transformation is given by $\underline{u}_2$, the eigenvector corresponding to the *smallest* eigenvalue. Hence,

$$\underline{y} = [0.77334, \ 0.63399]\underline{\beta}$$

All signals in the new one-dimensional features space are the projections of $\underline{\beta}_i$ on the line along the eigenvector $\underline{u}_2$. This projection is shown in Figure 13. This figure clearly demonstrates that the clustering in the reduced, one-dimensional space was preserved. Note that projections on the eigenvector $\underline{u}_1$ (corresponding to the largest eigenvalue) do not preserve

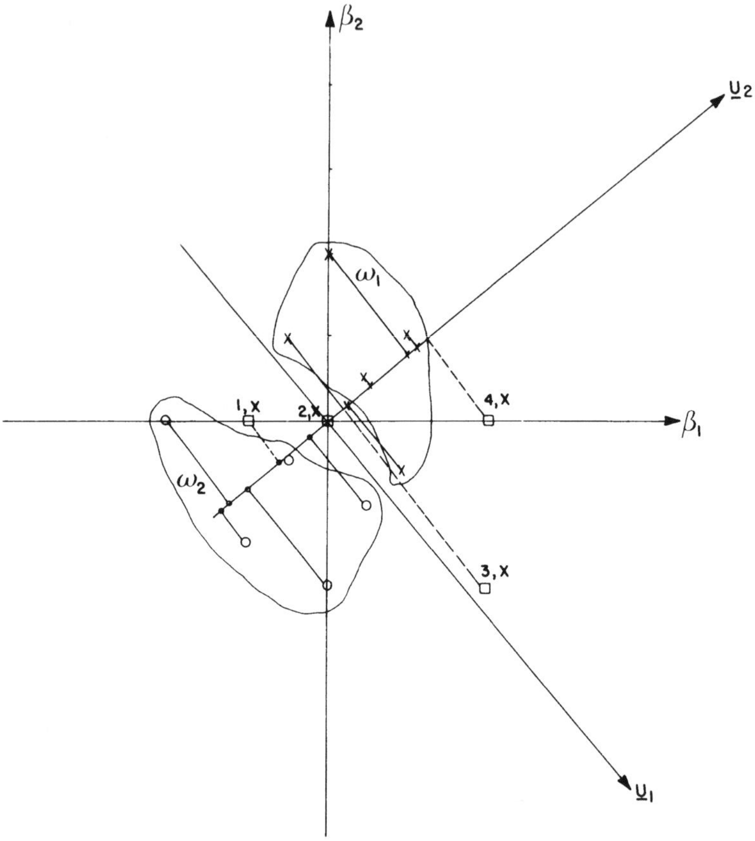

FIGURE 13. Compression and classification by means of entropy criteria.

the discrimination between classes. For classification we need to find a decision surface (a threshold number in the one-dimensional case). If, for example, we take y = 0 to be the threshold, we see that all the training data are classified correctly. The unknown data is classified as follows:

$$\underline{\beta}_{1,x} \in w_2 \quad \underline{\beta}_{2,x} \in \text{Undefined} \quad \underline{\beta}_{3,x} \in w_1 \quad \underline{\beta}_{4,x} \in w_1$$

Comparing these results with Examples 3.1, 3.3, and 3.4, we note that the various classifiers differ in the classification of signals $\underline{\beta}_{2;x}$ and $\underline{\beta}_{3,x}$ which are indeed borderline cases.

3. Maximization of Entropy

We shall consider now a transformation similar to Equation 3.67, but rather than requiring the minimization of the dispersion, we shall require that maximum information be transformed. This means we want to maximize the entropy, H(y) (rather than minimize it as was done in the previous section). Let us assume that the probabilities involved are normal, with identical covariance matrix, Σ_β, for all classes. The transformation T that maximizes the entropy will be the transformation, the columns of which are d eigenvectors of Σ_β, corresponding to the *largest* eigenvalues.

Example 3.6

Consider the data given in Example 3.1. Find the transformation into the one-dimensional space that will preserve maximum information in the sense of the entropy. Clearly the transformation is the vector $\underline{u}_1$ of Example 3.5. The vector $\underline{u}_1$ is shown in Figure 13. Note that the projections of the signals onto $\underline{u}_1$ completely destroy the discrimination between the classes.

IV. FISHER'S LINEAR DISCRIMINANT

Fisher's linear discriminant is a transformation that reduces the dimensionality of the features vector from n into $d = M - 1$ (where M is the number of classes involved), while optimally preserving the separability between classes. The compression ratio is dictated by the original dimensionality and the number of classes and cannot be chosen at will as was the case, for example, with the minimum entropy criterion.

The idea behind the Fisher's linear discriminant is the projection of the n dimensional features vectors onto a lower dimensional surface. The surface is to be chosen such that separation between classes is kept as much as possible.

Refer to Figure 13 and note that if we have chosen (as the minimum entropy method did) to project the data onto $\underline{u}_2$, the projected classes remain separate. Any other one-dimensional surface would have yielded projected classes which are less separated or even intermingled as is clearly the case if $\underline{u}_1$ is chosen.

In order to find the optimal surface to project onto, a measure of separability is required. Optimality is then understood in the sense of maximizing separability. The criterion chosen in the discussion presented in the last section was that of minimum entropy. Here we take a different approach.

Consider first the two-class problem. Suppose we have N known samples, $\underline{\beta}_i$, N_1 of which belong to w_1 and N_2 of which belong to w_2. Consider y_i to be a linear combination of the features $\underline{\beta}_i$:

$$y_i = \underline{\rho}^T\underline{\beta}_i \tag{3.71}$$

The n dimensional vector, $\underline{\rho}$, can be considered a line in the n dimensional space; then y_i is the projection of $\underline{\beta}_i$ on this line (scaled by $\|\underline{\rho}\|$). Let $\underline{\hat{\mu}}_i$ be the mean of the N_i samples of class w_i in the n dimensional space:

$$\underline{\hat{\mu}}_i = \frac{1}{N_i}\sum_{\underline{\beta}\in w_i} \underline{\beta}_i \tag{3.72}$$

and the mean of the projected points y_i on the line $\underline{\rho}$, $\tilde{\mu}_i$, is the projection of $\hat{\mu}_i$:

$$\tilde{\mu}_i = \frac{1}{N_i}\sum_{y_i\in w_i} y_i = \frac{1}{N_i}\sum_{\underline{\beta}_i\in w_i} \underline{\rho}^T\underline{\beta}_i = \underline{\rho}^T\underline{\hat{\mu}}_i \tag{3.73}$$

The separation of the means on $\underline{\rho}$ is given by

$$|\tilde{\mu}_1 - \tilde{\mu}_2| = |\underline{\rho}^T(\underline{\hat{\mu}}_1 - \underline{\hat{\mu}}_2)| \tag{3.74}$$

It can be made as large as required by scaling $\underline{\rho}$. This by itself is, of course, meaningless since the separation of the two classes must include the variance of the samples.

Define the $n \times n$ scatter matrix, W_i, of the *ith* class as:

$$W_i = \sum_{\underline{\beta}\in w_i} (\underline{\beta} - \underline{\hat{\mu}}_i)(\underline{\beta} - \underline{\hat{\mu}}_i)^T$$

$$i = 1,2 \tag{3.75}$$

W_i is the estimation of the covariance of the *ith* class in the n-dimensional feature space. It represents a measure of the dispersion of the signals belonging to w_i. The within-class scatter matrix, W, is defined as:

$$W = W_1 + W_2 \tag{3.76}$$

The one-dimensional scatter "matrix" of the projections y_i is similarly given by:

$$\hat{\sigma}_i^2 = \sum_{y \in w_i} (y - \tilde{\mu}_i)^2 = \sum_{\underline{\beta} \in w_i} (\underline{\rho}^T \underline{\beta} - \underline{\rho}^T \underline{\hat{\mu}}_i)^2$$
$$= \sum_{\underline{\beta} \in w_i} \underline{\rho}^T (\underline{\beta} - \underline{\hat{\mu}}_i)(\underline{\beta} - \underline{\hat{\mu}}_i)^T \underline{\rho} = \underline{\rho}^T W_i \underline{\rho} \tag{3.77}$$

and

$$\hat{\sigma}_1^2 + \hat{\sigma}_2^2 = \underline{\rho}^T W \underline{\rho} \tag{3.78}$$

Consider now the variance between the means of the two classes. Denote the matrix, B, the "between class scatter matrix", in the original n-dimensional features space:

$$B = (\underline{\hat{\mu}}_1 - \underline{\hat{\mu}}_2)(\underline{\hat{\mu}}_1 - \underline{\hat{\mu}}_2)^T \tag{3.79}$$

This matrix represents the dispersion between the means of the various classes. Also the variance of the means in the one-dimensional projection is

$$(\tilde{\mu}_1 - \tilde{\mu}_2)^2 = (\underline{\rho}^T \underline{\hat{\mu}}_1 - \underline{\rho}^T \underline{\hat{\mu}}_2)^2 = \underline{\rho}^T (\underline{\hat{\mu}}_1 - \underline{\hat{\mu}}_2)(\underline{\hat{\mu}}_1 - \underline{\hat{\mu}}_2)^T \underline{\rho} = \underline{\rho}^T B \underline{\rho} \tag{3.80}$$

Note that for every n-dimensional vector $\underline{\upsilon}$, we have from Equation 3.79:

$$B\underline{\upsilon} = (\underline{\hat{\mu}}_1 - \underline{\hat{\mu}}_2)(\underline{\hat{\mu}}_1 - \underline{\hat{\mu}}_2)^T \underline{\upsilon} \tag{3.81}$$

Since $(\underline{\hat{\mu}}_1 - \underline{\hat{\mu}}_2)^T \underline{\upsilon}$ is a scalar denoting the projection of $\underline{\upsilon}$ of $(\underline{\hat{\mu}}_1 - \underline{\hat{\mu}}_2)$, we conclude that $B\underline{\upsilon}$ is always a vector in the direction of $(\underline{\hat{\mu}}_1 - \underline{\hat{\mu}}_2)$.

A criterion of separation can now be formulated in terms of the new scatter matrices. For good separation we require that the variance of the populations of each class be small. Hence a good separation measure, $J(\underline{\rho})$, is

$$J(\underline{\rho}) = \frac{\underline{\rho}^T B \underline{\rho}}{\underline{\rho}^T W \underline{\rho}} \tag{3.82}$$

which is known as the Rayleigh quotient. If W is nonsingular, the maximization of Equation 3.82 is given by the eigenvalue equation:

$$W^{-1} B \underline{\rho} = \lambda \underline{\rho} \tag{3.83}$$

where the optimal weighting vectors, $\underline{\rho}$, are the eigenvectors of $W^{-1}B$. In this case, however, we need not solve for the eigenvectors. Recall that $B\underline{\rho}$ is a vector in the direction of $(\underline{\hat{\mu}}_1 - \underline{\hat{\mu}}_2)$ let its length be λ. We can do this without loss of generality since the length of the required vector is of no importance; its direction is what we look for. Hence, we get

$$\underline{\rho} = W^{-1}(\underline{\hat{\mu}}_1 - \underline{\hat{\mu}}_2) \tag{3.84}$$

The line along the vector $\underline{\rho}$ given by Equation 3.84 is the optimal line in the sense that the projections of w_1 and w_2 on it will have the maximum ratio of "between" to "within" class scatter. The classification problem has now been reduced to that of finding a decision surface (threshold number) on the line $\underline{\rho}$ to discriminate between the projections of w_1 and w_2.

Example 3.7

Consider again the data given in Example 3.1 (Figure 4). It is desired to reduce the dimensions of the feature vector, $\underline{\beta}$, to one dimension using the Fisher's discriminant. We thus have to calculate the transformation vector, $\underline{\rho}$, of Equation 3.71. The direction of the vector is given by Equation 3.84. For this example (see Example 3.1),

$$\underline{\rho} = \hat{\Sigma}^{-1}(\hat{\underline{\mu}}_1 - \hat{\underline{\mu}}_2) = 5.353\begin{vmatrix} 1.000 \\ 1.009 \end{vmatrix} \approx 5.353\begin{vmatrix} 1 \\ 1 \end{vmatrix}$$

which yields a transformation very close to $\underline{u}_2$ of Example 3.5 (Figure 13). The classification results will be similar to those of Example 3.5 with signal $\underline{\beta}_{3,x}$ undefined.

Consider now the general case where M classes are present. The within class scatter matrix (Equation 3.76) will become:

$$W = \sum_{i=1}^{M} W_i \tag{3.85}$$

For the generalized between matrix, we calculate the mean of classes:

$$\hat{\underline{\mu}} = \frac{1}{N}\sum_{i=1}^{M} N_i\hat{\underline{\mu}}_i \tag{3.86}$$

and the between scatter matrix,

$$B = \sum_{i=1}^{M} N_i(\hat{\underline{\mu}}_i - \hat{\underline{\mu}})(\hat{\underline{\mu}}_i - \hat{\underline{\mu}})^T \tag{3.87}$$

We shall calculate M − 1 discriminant functions.

The reduction in dimensionality will thus be from the original n to M − 1. We of course assume that n > M − 1. The discriminant functions are given by the projections of the signal samples $\underline{\beta}$ onto the lines $\underline{\rho}_i$:

$$y_i = \underline{\rho}_i^T\underline{\beta} \qquad i = 1,2,\ldots,M - 1 \tag{3.88}$$

The M − 1 equations can be written in a matrix form using the M − 1 dimensional vector $\underline{y}^T = [y_1,y_2, \ldots ,y_{M-1}$ and n × (M − 1) matrix T whose columns are the weighting vectors $\underline{\rho}_i$.

$$\underline{y} = T^T\underline{\beta} \tag{3.89}$$

Equation 3.89 gives the transformation onto the M − 1 space. The optimal transformation matrix T is to be calculated.

The within class scatter matrix in the reduced M − 1 space is denoted by W_y and the between class scatter matrix in that space is denoted by B_y. Similar to the two-classes case, we have

$$W_y = T^TWT$$
$$B_y = T^TBT \tag{3.90}$$

We need now a criterion for separability. The ratio of scalar measures used in the reduced, one-dimensional case cannot be used here since ratio of matrices is not defined. We could have used the criterion $tr(W_y^{-1}B_y)$ using the same logic as before. Another criterion can be the ratio of determinants:

$$J(T) = \frac{|T^TBT|}{|T^TWT|} \tag{3.91}$$

The matrix, T, that maximizes[1] Equation 3.91 is the one whose columns $\underline{\rho}_i$ are the solution of the equation:

$$B\underline{\rho}_i = \lambda_i W\underline{\rho}_i$$
$$i = 1,2,\ldots,M - 1 \tag{3.92A}$$

which can be solved either by inverting W and solving the eigenvalue problem:

$$W^{-1}B\underline{\rho}_i = \lambda_i\underline{\rho}_i \tag{3.92B}$$

or by solving[1] for the eigenvalues λ_i from

$$|B - \lambda_i W| = 0$$

and then solving $\underline{\rho}_i$ from

$$(B - \lambda_i W)\underline{\rho}_i = 0 \tag{3.92C}$$

The transformation (Equation 3.89) that transforms the n-dimensional features vector $\underline{\beta}$ into a reduced, M − 1, dimensional vector $\underline{y}$, while maximizing Equation 3.91, is given by Equation 3.92. The optimal transformation is thus the matrix, T, whose columns, $\underline{\rho}_i$, i = 1,2, . . . ,M − 1, are the eigenvectors of $W^{-1}B$. The Fisher's discriminant method is therefore useful for signal compression when classification in the reduced space is required.

V. KARHUNEN-LOEVE EXPANSIONS (KLE)

A. Introduction

The problem of dimensionality reduction is well known in statistics and in communication theory. A variety of methods have been developed, employing linear transformation, that transform the original feature space into a lower order space while optimizing some given performance index. Two classical methods in statistics are the principal components analysis[3-5,40] (PCA), known in the communication theory literature as Karhunen-Loeve Expansion

(KLE), and factor analysis (FA). The PCA optimizes the variance of the features while FA optimizes the correlations among the features.

The KLE has been used to extract important features for representing sample signals taken from a given distribution. To this end the method is well suited for signal compression. In classification, however, we wish to represent the features which possess the maximum discriminatory information between given classes and not to faithfully represent each class by itself. There may be indeed cases where two classes may share the same (or similar) important features, but also have some different features (which may be less important in terms of representing each class). If we reduce the dimensions of the classes by keeping only the important features, we lose all discriminatory information. It has been shown[40] that if a certain transformation is applied to the data, *prior* to KLE, discrimination is preserved.

The KLE applied to a vector, representing the time samples, can be extended to include several signals. We arrange the vectors representing a group of signals into a matrix form and try to represent this "data" matrix in lower dimension, namely, by means of a lower rank matrix. This extension to the KLE (PCA) is known as singular value decomposition (SVD).

Principal components analysis (PCA, KLE) has been widely applied to biomedical signal processing.[14,20,30,32,43] SVD methods[45,47-50] have also been applied to the biomedical signal processing, in particular to ECG[51] and EEG processing.[47]

B. Karhunen-Loeve Transformation (KLT) — Principal Components Analysis (PCA)

Consider again the transformation of Equation 3.67. Assume that $d = n$, namely, the transformation from the n-dimensional space onto itself. We wish to choose transformation T with orthonormal vectors:

$$T^T = [\underline{\Phi}_1 \vdots \underline{\Phi}_2 \vdots , \ldots , \vdots \underline{\Phi}_n]$$

$$\underline{\Phi}_i^T \underline{\Phi}_j = \begin{cases} 1 \ , \ i = j \\ 0 \ , \ i \neq j \end{cases} \tag{3.93}$$

Note that here T is a square $n \times n$ orthogonal matrix for which

$$T^{-1} = T^T \tag{3.94}$$

and

$$\underline{\beta} = T^T \underline{y} = [\underline{\Phi}_1 \vdots \underline{\Phi}_2 \vdots , \ldots , \vdots \underline{\Phi}_n] \underline{y} = \sum_{i=1}^{n} y_i \underline{\Phi}_i \tag{3.95}$$

where y_i are the elements of $\underline{y}$.

The transformation from the original n-dimensional features vector, $\underline{\beta}$, into the new (n dimensional) features vector, $\underline{y}$, is given in Equation 3.95 as an expansion of $\underline{\beta}$ by means of a set of n orthonormal vectors, $\underline{\Phi}_i$.

We wish now to reduce the dimensionality of $\underline{y}$ from n to some $d < n$. We shall select d parts of Equation 3.95 and replace the rest $(n - d)$ by preselected constants b_i. We shall then reconstruct the original features vector by

$$\hat{\underline{\beta}}(d) = \sum_{i=1}^{d} y_i \underline{\Phi}_i + \sum_{i=d+1}^{n} b_i \underline{\Phi}_i \tag{3.96}$$

since b_i, $i = d + 1, \ldots, n$, are preselected constants, the vector $\underline{y}$ describing the signal is d dimensional. The reconstruction error, $\Delta\underline{\beta}(d)$, is

$$\Delta\underline{\beta}(d) = \underline{\beta} - \hat{\underline{\beta}}(d) = \underline{\beta} - \sum_{i=1}^{d} y_i\underline{\Phi}_i - \sum_{i=d+1}^{n} b_i\underline{\Phi}_i = \sum_{i=d+1}^{n} (y_i - b_i)\underline{\Phi}_i \quad (3.97)$$

The mean square reconstruction error, $\underline{\epsilon}(d)$, is given from Equations 3.97 and 3.93 by

$$\underline{\epsilon}(d) = E\{\Delta\underline{\beta}(d)^T\Delta\underline{\beta}(d)\} = E\left\{\sum_{i=d+1}^{n}\sum_{j=d+1}^{n}(y_i - b_i)(y_j - b_j)\underline{\Phi}_i^T\underline{\Phi}_j\right\}$$

$$= \sum_{i=d+1}^{n} E\{(y_i - b_i)^2\} \quad (3.98)$$

We shall choose b_i and $\underline{\Phi}_i$ that minimize the mean square error of Equation 3.98. To get the optimal b_i's, derive Equation 3.98 with respect to b_i:

$$\frac{\partial\underline{\epsilon}(d)}{\partial b_i} = -2(E\{y_i\} - b_i) = 0$$

$$b_i = E\{y_i\} = \underline{\Phi}_i^T E\{\underline{\beta}\} \quad (3.99)$$

Equation 3.98 can be rewritten using Equation 3.99:

$$\underline{\epsilon}(d) = \sum_{i=d+1}^{n} E\{(y_i - b_i)(y_i - b_i)^T\}$$

$$= \sum_{i=d+1}^{n} \underline{\Phi}_i^T E\{(\underline{\beta} - E\{\underline{\beta}\})(\underline{\beta} - E\{\underline{\beta}\})^T\}\underline{\Phi}_i$$

$$= \sum_{i=d+1}^{n} \underline{\Phi}_i^T \Sigma_\beta \underline{\Phi}_i \quad (3.100)$$

The optimal vectors, $\underline{\Phi}_i$, and thus the optimal transformation T, are given by the minimization of Equation 3.100 with the constraint $\underline{\Phi}_i^T\underline{\Phi}_i = 1$. Using the Lagrange multipliers, λ_i, the constraint minimization result is[3]

$$\Sigma_\beta\underline{\Phi}_i = \lambda_i\underline{\Phi}_i \quad (3.101)$$

The solution of Equation 3.101 provides the optimal $\underline{\Phi}_i$ which are the eigenvectors of the original covariance matrix Σ_β, corresponding to the eigenvalues λ_i. Substituting Equation 3.101 into Equation 3.100 yields the minimum square error:

$$\underline{\epsilon}(d)_{min} = \sum_{i=d+1}^{n} \lambda_i \quad (3.102)$$

Note that since $\underline{\epsilon}(d)$ is positive, the eigenvalues are nonnegative. For data compression we shall choose d columns in the transformation, T, in such a way that the error (Equation

3.102) is minimal. We shall choose the d eigenvectors, corresponding to the largest eigenvalues for the transformation and delete the rest (n − d) eigenvectors. The error (Equation 3.102) will then consist of the sum of (n − d) smallest eigenvalues.

We can arrange the eigenvalues such that:

$$\lambda_1 \geq \lambda_2 \geq \lambda_3 \geq ,\ldots \geq \lambda_n \geq 0$$

Then the required transformation matrix, T^T, consists of the columns, $\underline{\Phi}_i$, i = 1,2, . . . ,d. Note also that

$$\Sigma_y = T\Sigma_\beta T^{-1} = T\Sigma_\beta T^T = \text{diag}(\lambda_1,\lambda_2,\ldots,\lambda_d) \quad (3.103)$$

since T is the modal matrix of Σ_β. The covariance matrix of $\underline{y}$ is diagonal which means that the features y_i are uncorrelated.

The calculation of the eigenvectors of the matrix, Σ_β, is not an easy task. Several algorithms for this calculation have been suggested.[3,44]

Some results of KLE as applied to biomedical signals are shown in Figure 14. The KLE as presented in this section is very effective for signal compression when the application is effective storing or reconstruction. This is obvious since it is optimal in the sense of minimum square of reconstruction error. It is, however, less attractive when the goal is classification.

Fukunaga and Koontz[40] have suggested a method whereby a modified KLE can be used for two-class discrimination. Their method is not optimal and indeed a counter example has been presented[52] that shows poor discriminant behavior of the method.

C. Singular Value Decomposition (SVD)

The KLE was designed to expand a signal $\underline{\beta}$. We can group all L given signals belonging to one class into an n × L matrix F,

$$F = [\underline{\beta}_1 \vdots \underline{\beta}_2 \vdots ,\ldots \vdots \underline{\beta}_L] \quad (3.104)$$

and expand the matrix. If the rank of the matrix B is k, it can be expressed as[54]

$$F = US_FV^T \quad (3.105)$$

where the n × n matrix U and L × L matrix V are unitary matrices ($V^T = V^{-1}$) and the rectangular n × L matrix, S_F, is a diagonal matrix with real nonnegative diagonal elements, s_i, known as the singular values. It is obvious that k is less than or equal to the minimum of n and L.

The singular values are conventionally ordered in decreasing order $s_1 \geq s_2 \geq , \ldots , s_k \geq 0$, with the largest one, s_1, in the upper left hand corner of S_F. The singular values are the nonnegative square roots of the eigenvalues of FF^T and F^TF. The n × 1 dimensional vector $\underline{u}_i$, i = 1,2, . . . ,n (the columns of U), and the L × 1 dimensional vectors, $\underline{v}_i$, i = 1,2, . . . ,L (the columns of V), are the orthonormal eigenvectors of FF^T and F^TF, respectively.

The representation given in Equation 3.105 states that any k rank matrix can be expressed by means of a rectangular diagonal matrix, multiplied from right and left by unitary matrices. Equation 3.105 can also be written as

$$F = \sum_{i=1}^{k} s_i \underline{u}_i \underline{v}_i^T \quad (3.106)$$

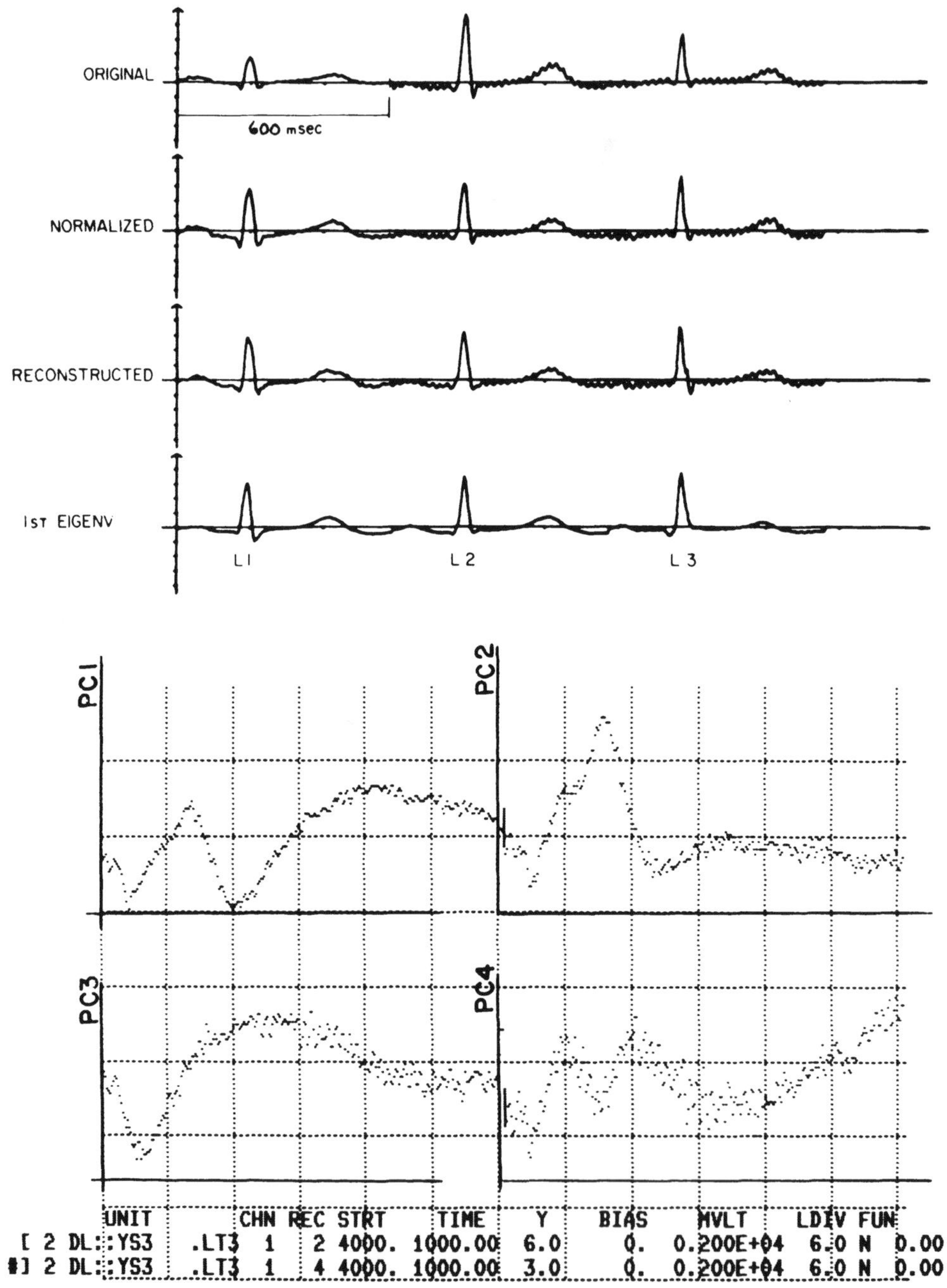

FIGURE 14. Karhunen-Loeve analysis, most significant eigenvectors of (A) ECG signal and (B) pain evoked potentials.

in Equation 3.106 the matrix F is expressed in terms of the sum of k matrices with rank one. The matrices $\underline{u}_i\underline{v}_i^T$ are called singular planes or eigenplanes.

Since the eigenvectors are orthonormal, it follows from Equation 3.106 that

$$\underline{u}_i = \frac{1}{s_i} F\underline{v}_i \qquad (3.107A)$$

$$\underline{\upsilon}_i = \frac{1}{s_i} F^T \underline{u}_i \tag{3.107B}$$

Recall that the L columns of the features matrix F are the signals $\underline{\beta}_j$, j = 1,2, . . . ,L. Each vector then can be expressed from Equation 3.106 as

$$\underline{\beta}_j = \sum_{i=1}^{k} s_i \upsilon_{i,j} \underline{\mu}_i$$

$$j = 1,2,\ldots,L \tag{3.108}$$

where the elements of $\underline{\upsilon}_i$ are

$$\underline{\upsilon}_i^T = [\upsilon_{i,1}, \upsilon_{i,2}, \ldots, \upsilon_{i,L}]$$

We can define the coefficients $C_{i,j}$:

$$C_{i,j} = s_i \upsilon_{i,j}$$

$$i = 1,2,\ldots,k$$

$$j = 1,2,\ldots,L$$

and rewrite Equation 3.108 by

$$\underline{\beta}_j = C_{1,j}\underline{u}_1 + C_{2,j}\underline{u}_2 + \ldots + C_{k,j}\underline{u}_k \tag{3.109}$$

In Equation 3.108 each signal $\underline{\beta}_j$ is expressed in terms of eigenvectors of FF^T. The SVD is thus, in principle, equivalent to the PCA.

We want to use the SVD for compression. Consider now the expansion of Equation 3.106 with summation index running until $d \leq k$. Note that since we have arranged the singular values in decreasing order, the reduced expansion will include the d largest values. If we thus denote the estimate of F by $\hat{F}$:

$$\hat{F} = \sum_{i=1}^{d} s_i \underline{u}_i \underline{\upsilon}_i^T = U\hat{S}_F V^T$$

$$d \leq k \tag{3.110}$$

Then, for any matrix C, of rank d, the following holds true:

$$\|F - \hat{F}\| \leq \|F - C\| \tag{3.111}$$

Here the norm is in the sense of the sum of squares of all entries:

$$\|A\|^2 = \mathrm{tr}[(A)^T(A)] = \sum_{i,j} a_{i,j}^2 \tag{3.112}$$

In Equation 3.110 the singular values are the d largest ones and $\hat{S}_F$ is obtained from S_F by setting to zero all but the d largest singular values. Equation 3.111 states that the estimated reduced matrix $\hat{F}$ is the best least squares estimates of F by rank d matrices.

Analogous to PCA, it can be shown here that the estimation error (residual error), expressed in terms of the norm of Equation 3.111, equals the sum of squares of the discarded singular values:

$$\|F - \hat{F}\|^2 = \sum_{i=d+1}^{k} s_i^2 \tag{3.113}$$

The reason for choosing d largest singular values for the expansion is clear from Equation 3.113.

Methods for computing the singular values from the eigenvalues of FF^T require heavy computation load and are time consuming. In addition, conventional methods yield all k singular values where indeed only the d largest ones are required. A method for the computation of the singular values, one at a time, beginning with the largest one, has been suggested;[50] it is known as the power method. The computation can be stopped when the already acquired singular values yield residual error (Equation 3.113) below a certain threshold. The method is especially attractive in cases where the data matrix is large, but its rank is low. The computation method is briefly presented here. Note also that this method operates on the data matrix F directly and not on the correlation matrix FF^T.

The computation is based on the solution of the two equations (Equation 3.107):

$$s\underline{u} = F\underline{\upsilon} \tag{3.114A}$$

$$s\underline{\upsilon} = F^T\underline{u} \tag{3.114B}$$

Using an arbitrary starting vector $\underline{\upsilon}^{(0)}$, we form the following iterative solution:

$$\underline{u}^{(k+1)} = \frac{F\underline{\upsilon}^{(k)}}{\|F\underline{\upsilon}^{(k)}\|} \tag{3.115A}$$

$$\underline{\upsilon}^{(k+1)} = \frac{F^T\underline{u}^{(k+1)}}{\|F^T\underline{u}^{(k+1)}\|} \tag{3.115B}$$

We continue with the iteration until

$$\underline{u}^{(k+1)} - \underline{u}^{(k)} \leqslant \underline{\epsilon} \tag{3.116}$$

where $\underline{\epsilon}$ is some predetermined stopping vector. The first (largest) singular value, s_1, is then estimated by taking the norm of Equation 3.114B and recalling that the length of $\underline{\upsilon}$ is unity:

$$\hat{s}_1 = \|F^T\underline{u}^{(k+1)}\| \tag{3.117A}$$

and the orthonormal vectors are

$$\hat{\underline{u}}_1 = \underline{u}^{(k+1)}$$

$$\hat{\underline{\upsilon}}_1 = \underline{\upsilon}^{(k+1)} \tag{3.117B}$$

for the last k. To obtain the next singular value and eigenvectors, the estimated singular plane is removed from F by

$$F^{(1)} = F - \hat{s}_1\hat{\underline{u}}_1\hat{\underline{\upsilon}}_1^T \tag{3.118}$$

and the same iterations (Equations 3.115 to 3.118) are repeated for $F^{(1)}$ to get $\hat{s}_2$, $\hat{\underline{u}}_2$, $\hat{\underline{\upsilon}}_2$ and so on until $\hat{s}_d$, $\hat{u}_d$, $\hat{v}_d$. The last (d*th*) singular value to be calculated is determined by the threshold ϵ placed on the residual error:

$$\|F - \sum_{i=1}^{d} \hat{s}_i \underline{\hat{u}}_i \underline{\hat{v}}_i^T\| \leq \epsilon \tag{3.119}$$

The convergence properties of the algorithm are discussed by Shlien.[50] The convergence rate depends on the ratio of the adjacent highest eigenvalues, thus becoming very slow when this ratio is close to one. An algorithm that improves convergence in these critical cases has been suggested.[55]

Example 3.8

Consider again the data given in Example 3.1. To simplify the calculations we shall regard only the first three signals of w_1. We represent these signals by means of the matrix F:

$$F = [\underline{\beta}_{1,1} \vdots \underline{\beta}_{2,1} \vdots \underline{\beta}_{3,1}] = \begin{bmatrix} -0.5 & 0 & 1 \\ 1 & 2 & 1 \end{bmatrix}$$

The rank of the matrix is $k = 2$. The 2×2 correlation matrix FF^T is

$$FF^T = \begin{bmatrix} 1.25 & 0.5 \\ 0.5 & 6 \end{bmatrix}$$

The eigenvalues of FF^T are $\lambda_1 = 6.05206$ and $\lambda_2 = 1.19794$ with corresponding singular values $s_1 = \lambda_1^{1/2} = 2.46$ and $s_2 = \lambda_2^{1/2} = 1.0945$. The matrix U is

$$U = [\underline{u}_i \vdots \underline{u}_2] = \begin{bmatrix} 0.10356 & -0.99462 \\ 0.99462 & 0.10356 \end{bmatrix}$$

where $\underline{u}_1$ and $\underline{u}_2$ are the orthornormal eigenvectors of FF^T corresponding to λ_1 and λ_2.

The eigenvalues and eigenvectors of F^TF are

$$F^TF = \begin{bmatrix} 1.25 & 2 & 0.5 \\ 2 & 4 & 2 \\ 0.5 & 2 & 2 \end{bmatrix}$$

$$\lambda_1 = 6.05206, \lambda_2 = 1.19794, \lambda_3 = 0, \text{ and:}$$

$$\nu = \begin{bmatrix} 0.3832 & 0.5490 & -0.7428 \\ 0.8086 & 0.18924 & 0.5571 \\ 0.4464 & -0.81398 & -0.3714 \end{bmatrix}$$

The matrix F can now be expanded by

$$F = \sum_{j=1}^{2} s_i \underline{u}_i \underline{\nu}_i^T =$$

$$= \begin{bmatrix} 0.0976 & 0.206 & 0.11372 \\ 0.9376 & 1.9784 & 1.0922 \end{bmatrix}$$

$$+ \begin{bmatrix} -0.5976 & -0.206 & 0.8861 \\ 0.0622 & 0.02144 & -0.0922 \end{bmatrix}$$

$$= \begin{bmatrix} -0.5 & 0 & 1 \\ 1 & 2 & 2 \end{bmatrix}$$

If we desire to reduce the dimensions, we shall take only the first term in the expansion, namely, the projection on the first eigenplane:

$$\hat{F} = s_1 \underline{u}_1 \underline{\nu}_1^T = \begin{bmatrix} 0.0976 & 0.206 & 0.11372 \\ 0.9376 & 1.9784 & 1.0922 \end{bmatrix}$$

$\hat{F}$ is a matrix of rank 1. The residual error for the estimate is

$$\|F - \hat{F}\|^2 = \left\| \begin{matrix} -0.5976 & -0.206 & 0.8861 \\ 0.0622 & 0.02144 & -0.0922 \end{matrix} \right\| = 1.19756 \cong s_2^2 = \lambda_2$$

Let us repeat this example with Shlien's power method. Choose the initial unit length vector $\underline{\upsilon}^{(0)}$ to be

$$\underline{\nu}^{(0)} = [3^{-1/2}\ 3^{-1/2}\ 3^{-1/2}]$$

Using the iterative equations we get

$$\underline{u}^{(1)} = [0.1241,\quad 0.9923]^T$$

$$\underline{\nu}^{(1)} = [0.3764, \quad 0.829, \quad 0.4517]^T$$

$$\underline{u}^{(2)} = [0.1076, \quad 0.9942]^T$$

$$\underline{\nu}^{(2)} = [0.3823, \quad 0.8083, \quad 0.4478]^T$$

$$\underline{u}^{(3)} = [0.1043, \quad 0.9945]^T$$

$$\underline{\nu}^{(3)} = [0.3831, \quad 0.8085, \quad 0.4467]^T$$

We note that after the third iteration we have

$$|\underline{u}^{(3)} - \underline{u}^{(2)}| = |[-0.0033, \quad 0.0003]^T|$$

$$|\underline{\nu}^{(3)} - \underline{\nu}^{(2)}| = |[0.0008, \quad 0.0002, \quad -0.0011]^T|$$

If we consider this to be converged, we shall have

$$\underline{\hat{u}}_1 = \underline{u}^{(3)}, \ \underline{\hat{\nu}}_1 = \underline{\nu}^{(3)} \text{ and:}$$

$$\hat{s}_1 = ||B^T\underline{\hat{u}}_1|| = 2.474 \approx 2.46 = s_1$$

$$\hat{s}_2 = ||B\underline{\hat{\nu}}_1|| = 1.094 \approx 1.0945 = s_2$$

Expressing the matrix F by its reduced, rank one matrix, with largest singular value yields:

$$\hat{F} = 2.46\ \underline{\hat{u}}_1\underline{\hat{\nu}}_1^T = \begin{vmatrix} 0.0984 & 0.2066 & 0.1146 \\ 0.937 & 1.978 & 1.0928 \end{vmatrix}$$

which is very close to the reduced matrix calculated by the direct method.

VI. DIRECT FEATURE SELECTION AND ORDERING

A. Introduction

In classification problems with two or more classes, it is often required to choose a subset of d best features out of the given n or to arrange the features in order of importance. To do this we require a measure of class separability. The optimal measure of features effectiveness is the probability of error. In practice one can use the training data and use the percentage of classification error as a measure. This approach is often used; it is, however, experimental and requires a relatively large training data.

Scatter matrices can be used to form a separability criterion. Recall that the within class scatter matrix, W (Equations 3.75 and 3.85) is the covariance matrix of the features in a given class. The between class scatter matrix, B (Equation 3.87), is the covariance of the means of the classes.

A criterion for separability can be any criterion which is proportional to the between scatter matrix and also proportional to the inverse of the within scatter matrix. Maximization of such a criterion will ensure that while maximizing the "distance" between classes, we do not amplify (with the same rate) the scatter of the classes, thus causing no improvement to the separability. A criterion like this was used in Equations 3.82 and 3.91.

Several such criteria were suggested (e.g., see Fukunaga):[3]

$$\text{(a)}\quad J_1 = \text{tr}(W^{-1}B) \tag{3.120A}$$

$$\text{(b)}\quad J_2 = \ln|W^{-1}B| = \ln(|B|/|W|) \tag{3.120B}$$

$$\text{(c)}\quad J_3 = \text{tr}(B)/\text{tr}(W) \tag{3.120C}$$

These critera are relatively simple to use. However, they do not have a direct relationship to the probability of error. More complicated criteria that can be related to the error probability such as the Chernoff bound and Bhattacharyya distance are known.[3]

B. The Divergence

The divergence is a class separability measure similar to the Bhattacharyya distance; it is based on the likelihood ratio. Let the probability of getting a feature vector $\underline{\beta}$, given that it belongs to class w_i, be $p(\underline{\beta}|w_i)$. The ratio between these conditional probabilities for class w_i and class w_j yields information concerning the separability between these two classes. The logarithm of the ratio is usually taken with no loss to the concept.

The average discriminating information for class, w_i, with respect to w_j is given by averaging the log likelihood ratio over all w_j:

$$L_{i,j} = E\left\{\ln \frac{p(\underline{\beta}|w_i)}{p(\underline{\beta}|w_j)}\right\} = \int_{\underline{\beta}} p(\underline{\beta}|w_i)\ln \frac{p(\underline{\beta}|w_i)}{p(\underline{\beta}|w_j)}\, d\underline{\beta} \tag{3.121A}$$

Similarly, the average discriminating information of class w_j with respect to w_i is

$$L_{j,i} = \int_{\underline{\beta}} p(\underline{\beta}|w_j)\ln \frac{p(\underline{\beta}|w_j)}{p(\underline{\beta}|w_i)}\, d\underline{\beta} \tag{3.121B}$$

The total average information in Equations 3.121A and B is known as the divergence; let us denote it by $D_{i,j}$:

$$D_{i,j} = L_{i,j} + L_{j,i} = \int_{\underline{\beta}} (p(\underline{\beta}|w_i) - p(\underline{\beta}|w_j))\ln \frac{p(\underline{\beta}|w_i)}{p(\underline{\beta}|w_j)}\, d\underline{\beta} \tag{3.122}$$

For normal distributions with $\underline{\mu}_k$ and Σ_k, $k = i,j$, the expectations and covariance matrices, respectively, the divergence becomes

$$D_{i,j} = \frac{1}{2}(\underline{\mu}_i - \underline{\mu}_j)^T(\Sigma_i^{-1} + \Sigma_j^{-1})(\underline{\mu}_i - \underline{\mu}_j) + \frac{1}{2}\text{tr}(\Sigma_i^{-1}\Sigma_j + \Sigma_j^{-1}\Sigma_i - 2I) \tag{3.123}$$

When the two classes possess the same covariance matrix, $\Sigma_i = \Sigma_j = \Sigma$, the divergence becomes

$$D_{i,j} = (\underline{\mu}_i - \underline{\mu}_j)^T\,\Sigma^{-1}(\underline{\mu}_i - \underline{\mu}_j) \tag{3.124}$$

which is the Mahalanobis distance. Methods for selecting optimal features by the maximization of the divergence have been suggested.[4]

C. Dynamic Programming Methods

The problem of selecting the subset of best d features from a given set of n features ($d < n$) has a straightforward solution, that of exhaustive search. One can form all possible different combinations of d out of n features; for each one the measure of separability can be calculated and the best set selected. Unfortunately such an exhaustive search method is impossible for large n and d because of the amount of calculations required. For example, if $n = 40$ features are given and it is desired to find the best set of $d = 10$ features, 8.477×10^7 combinations have to be checked; clearly this is an impossible task. Several suboptimal methods have been suggested. "Search without replacement" or "knock-out" methods are simple to implement. Here we first search for the most effective single feature. The one selected is to be included in the final set. Out of the $n - 1$ features remaining, we look again for the single most effective feature that together with the feature chosen previously will be the best set, in the sense of maximizing the separability criterion. This procedure is continued until the set of d features is determined. The set thus selected is not necessarily the optimal set. The algorithm is, however, computationally attractive since it required only $d(n - (d - 1)/2)$ searches. For the case described before where $n = 40$ and $d = 10$, the number of searches is only 355.

An algorithm[56,57] based on dynamic programming has been applied to the problem of suboptimal selection of features. In this method we start with a single feature and find another feature that together form the optimal set of two features. We do this for all n features. To each set we attach the value of the criterion. At the end of the first step we have n sets of two features each with its criterion value. If $d = 2$ we choose the set with the highest criterion value. If $d > 2$ we proceed by adding to each one of the n sets a third feature in such a way that the set of three features maximizes the criterion. Again we attach the value of the criterion to the set. Note that now we have n sets of three features, each with its criterion value attached. We continue this way, with the n sets, until they include the required d features. We then choose the best set by finding the one that has the maximum value for the criterion.

An additional advantage provided by the algorithm is the fact that when selecting the best set of d features we also get all best set of p features, $p = 1,2, \ldots ,d$. This gives us information on the added value of the last feature at each step. Let us now formulate the algorithm. Recall that in each step we deal with n subsets. Consider the $(i - 1)$ step. We shall denote the n vectors selected up to this step by

$$\underline{\eta}^j_{i-1} = (\beta^j_1, \beta^j_2, \ldots, \beta^j_{i-1})^T$$
$$j = 1,2,\ldots,n \tag{3.125}$$

Here $\underline{\eta}^j_{i-1}$ is the jth, $(i - 1)$ dimensional vector and β^j_q, $q = 1,2, \ldots ,i - 1$, are the set of $(i - 1)$ features selected from the given n features ($\underline{\beta}$). In the ith step we shall increase the dimension of the n, $\underline{\eta}^j_{i-1}$ vectors by adding one feature (that is not already included in the vector) to each one of them, such that

$$\underline{\eta}^j_i = (\underline{\eta}^j_{i-1} \vdots \beta^j_i)^T$$
$$\beta^j_i \notin \underline{\eta}^j_{i-1}$$
$$j = 1,2,\ldots,n \tag{3.126}$$

The added feature β^j_i will be selected from the available $n - (i - 1)$ features such that the criterion will be maximized:

$$\lambda_i(\underline{\eta}_i^j) = \underset{\beta_i \notin \underline{\eta}_{i-1}^j}{\text{Max}} \{D_i(\underline{\eta}_{i-1}^j; \beta_i)\} \tag{3.127}$$

where $D_i(\underline{\eta})$ is the value of the divergence evaluated with the *ith* dimensional vector $\underline{\eta}$. λ_i denotes the maximum of the divergences. The algorithm proceeds up to the d*th* step. At that step the set with maximum divergence is selected:

$$\lambda_d(\underline{\eta}_d) = \underset{j}{\text{Max}}(\lambda_d(\underline{\eta}_d^j)) \tag{3.128}$$

The number of searches required by the algorithm[53] is n(d − 1)(n − d/2) which is substantially less than the exhaustive search (for large n and intermediate values of d). For the example chosen previously of n = 40 and d = 10, the dynamic programming algorithm requires 13,600 searches.

Example 3.9

The speech signal may be used to model the speaker's vocal tract. Since humans differ in the anatomy of the vocal tract and other speech producing systems, one can use speech features to identify or verify[58] speakers. In the example discussed here, 77 average features were extracted from frames of 15-sec speech. Each speaker was described by the 77-dimensional feature vector. It was desired to reduce the dimensionality to ten, using the dynamic programming method.

We shall not assume that the speakers possess identical covariance matrices. Hence, for the case of i-dimensional features we shall have the within class scatter matrices (Equation 3.85):

$$W_i^j = \Sigma(\underline{\beta}_i^j - \underline{\hat{\mu}}_i^j)(\underline{\beta}_i^j - \underline{\hat{\mu}}_i^j)^T; \quad j = 1,2,\ldots,NS$$

where NS is the number of speakers, $\underline{\beta}_i^j$ is the *ith* dimensional features vector of the *jth* speaker, $\underline{\hat{\mu}}_i^j$ is its estimated mean, and the summation is performed over all training samples available for the *jth* speaker. The between class scatter matrix (Equation 3.87) was defined:

$$B_i^j = \sum_{k=1}^{NS}(\underline{\hat{\mu}}_i^j - \underline{\hat{\mu}}_i^k)(\underline{\hat{\mu}}_i^j - \underline{\hat{\mu}}_i^k)^T; \quad j = 1,2,\ldots,NS$$

For each speaker, j, the divergence used was

$$D_i^j = \text{tr}((W_i^j)^{-1}B_i^j)$$

The original set of n = 77 features is given below:

AR coefficients	$\{a_i^V\}$, i = 1,2,...,10	$\{a_i^U\}$, i = 1,2,...,10
Correlations	$\{\rho_i^V\}$, i = 3,4,...,10	$\{\rho_i^U\}$, i = 3,4,...,10
PARCORs	$\{k_i^V\}$, i = 1,2,...,9	$\{k_i^U\}$, i = 1,2,...,9
Cepstrum	$\{C_i^V\}$, i = 2,3,...,10	$\{C_i^U\}$, i = 2,3,...,10
Prediction error	E_p^V	E_p^U
Log energy	LE^V	LE^U
Pitch	P	

Superscript V denotes average features for voiced (see Appendix A) segments and U denotes unvoiced segments. For a discussion of the features see Chapter 7, Volume I.

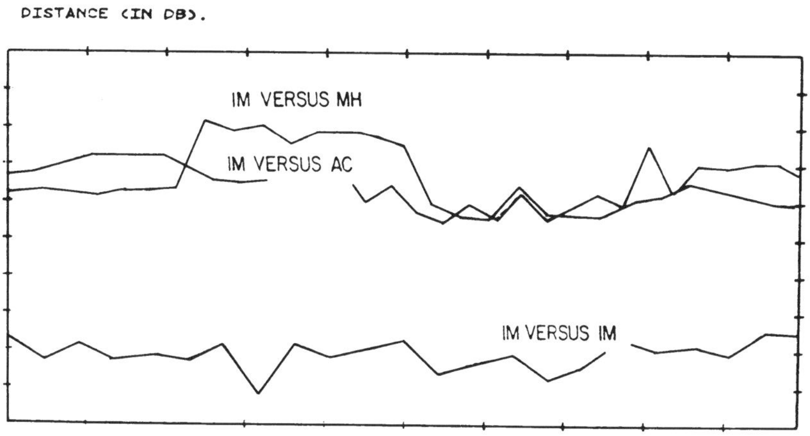

FIGURE 15. Distance between speakers in text independent speaker verification.

As an example, the results of the dynamic programming search for one speaker (AC) is shown here:

d	Features	Divergence
2	$[P, LE^V]$	250.67
3	$[P, LE^U, \rho_7^V]$	528.23
4	$[P, LE^U, \rho_8^V, \rho_{10}^V]$	684.06
5	$[P, LE^U, \rho_8^V, \rho_{10}^V, E_p^V]$	783.72
6	$[P, \rho_7^V, E_p^T, LE^U, \rho_4^V, k_2^V]$	941.52
7	$[P, LE^U, \rho_8^V, \rho_{10}^V, E_p^V, \rho_4^V, k_2^V]$	1096.6
8	$[P, \rho_7^V, E_p^V, k_1^V, LE^U, k_3^V, \rho_3^V, C_{10}^U]$	1325.9
9	$[P, \rho_7^V, E_p^V, k_1^V, LE^U, k_3^V, \rho_3^V, E_p^U, k_2^U]$	1563.9
10	$[P, \rho_7^V, E_p^V, k_1^V, LE^U, k_3^V, \rho_3^V, E_p^U, k_2^U, a_8^V]$	2036.9

Note that the pitch feature (P) was chosen in all subsets. The pitch is indeed known as an important feature for speaker identification. Note also that for low orders of features vectors an increase in dimensions changed the features (for example, note the suboptimal vectors of order 5 and 6). For larger orders, the main features did not change (e.g., see orders 9 and 10).

For actual speaker verification the Mahalanobis distance was used. Figure 15 shows the distances from segments of 15-sec speech of speaker (IM) to the templates of speakers IM, AC, and MH. The suboptimal feature vector of dimension 10 evaluated by the dynamic programming method was used.

VII. TIME WARPING

One of the fundamental problems that arises when comparing two patterns is that of time scaling. Up to now we have assumed that both pattern and template (reference) to be compared share the same time base. This is not always correct. The problem is especially severe in speech analysis.[60-64] It has been found that when a speaker utters the same word several times, he does so, in general, with different time bases for each utterance. Each word is spoken such that parts of it are uttered faster, and parts are uttered slower. The human brain, it seems, can easily overcome these differences and recognize the word. Machine recognition, however, finds this a severe difficulty. To overcome the problem, algorithms that map all patterns onto a common time base, thus allowing comparison, have been developed. These

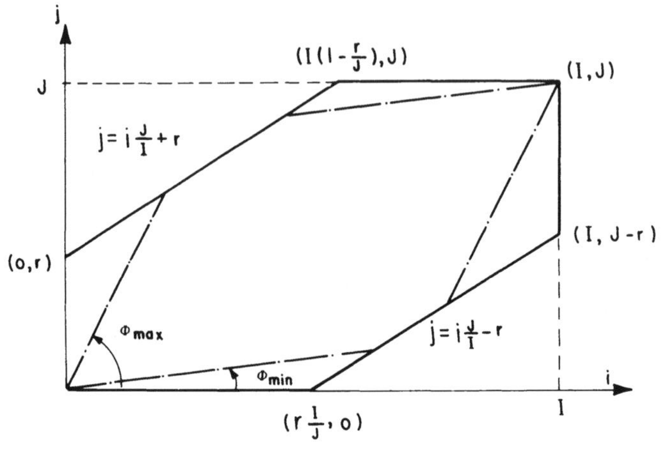

FIGURE 16. Time warping plane with search areas.

are known as "time warping" algorithms. The basic idea of time warping is depicted in Figure 16. Assume we are given two signals, $x(t_i)$, $x(t_j)$:

$$x(t_i)\ ,\ t_i \in (t_{is}, t_{if}) \tag{3.129A}$$

$$x(t_j)\ ,\ t_j \in (t_{js}, t_{jf}) \tag{3.129B}$$

each with its own time base, t_i and t_j. We assume that the beginning and end of each signal are known. These are denoted (t_{is}, t_{if}) and (t_{js}, t_{jf}), respectively. We shall consider the discrete case where both signals were sampled at the same rate. Assume also that the samples have been shifted such that both signals begin at sample $i = j = 1$. Without the loss of generality we have now:

$$x(i)\ ,\ i = 1,2,\ldots,I \tag{3.129C}$$

$$x(j)\ ,\ j = 1,2,\ldots,J \tag{3.129D}$$

If the two time bases were linearly related, the mapping function relating them was just $i = j \cdot I/J$. In general, however, the relation is nonlinear and one has to find the nonlinear time warping function. We shall make several assumptions on the warping function before actual calculations.

The warping function, W(k), is defined as a sequence of points:

$$c(1), c(2), \ldots, c(K)$$

where $c(k) = (i(k), j(k))$ is the matching of the point i(k) on the first time base and the point j(k) on the second time base. The warping, W(k), thus allows us to compare the appropriate parts of $x(t_i)$ with these of $x(t_j)$.

We shall impose a set of monotonic and continuity conditions on the warping function:[60]

$$0 \leq i(k) - i(k - 1) \leq p_i \tag{3.130A}$$

$$0 \leq j(k) - j(k - 1) \leq p_j \tag{3.130B}$$

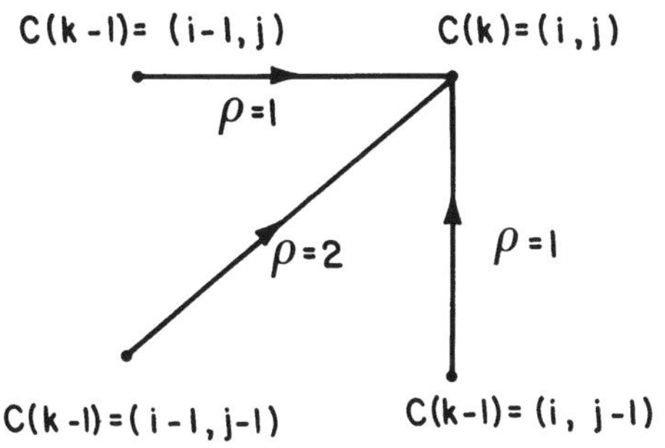

FIGURE 17. Constraints on the dynamic programming.

The left side inequality ensures increasing monotony; the right side inequality is a continuity condition that restricts the jumps in the warping. This restriction is important since large discontinuities can cause the elimination of parts of the signal. It has been suggested[60] to choose $p_i = p_j = 1$; we shall adapt this here. As a result of conditions (Equation 3.130), we have restricted the relations between two consecutive warping points c(k) and c(k − 1) to be

$$c(k - 1) = \begin{cases} (i(k)\ ,\ j(k) - 1) \\ (i(k) - 1\ ,\ j(k) - 1) \\ (i(k) - 1\ ,\ j(k) \end{cases} \tag{3.131}$$

Figure 17 depicts the meaning of the last equation. Due to the constraints, there are only three ways to get to the point c(k) = (i,j). These are given in Equation 3.131 and in Figure 17.

We also require boundary conditions. These will be defined as:

$$c(1) = (1,1)\ ,\ c(K) = (I,J) \tag{3.132}$$

By the boundary condition, we mean that we match the beginning and end of the signals. This is not always a good condition to impose since we may not have the endpoints of the signals accurately.

The warping function will be estimated by some type of search algorithm. We would like to limit the area over which the search is performed. We shall restrict the search window[62] to:

$$|i - j \cdot I/J| \leq \gamma \tag{3.133}$$

where γ is some given constant. The last condition limits the window to an area between lines parallel to the line j = iJ/I (see area bounded by solid lines in Figure 16).

Constraints on the slope can also be imposed. If we impose such conditions that limit the maximum allowable slope, ϕ_{max}, and minimum slope, ϕ_{min}, of the warping function, we end up with a parallelogram search window (see area bounded by broken lines in Figure 16).

We shall now proceed with the dynamic programming search. We recall now that the signals are represented, at each point, by their feature vectors, $\underline{\beta}_i(k)$ and $\underline{\beta}_j(k)$. Here $\underline{\beta}_i(k)$ denotes the feature vector of the signal $x(t_i)$ at the time i(k) with similar denotation for $\underline{\beta}_j(k)$.

Define a distance measure between the two feature vectors by

$$d(c(k)) = d(i(k),j(k)) = \|\underline{\beta}_i(k) - \underline{\beta}_j(k)\| \tag{3.134}$$

We will search for the warping function that will minimize a performance index, $D(x(t_i),x(t_j))$. We shall use the normalized average weighted distance as the performance index; hence,

$$D(x(t_i),x(t_j)) = \underset{W}{\mathrm{Min}}\left[\frac{\sum_{k=1}^{K} d(c(k))\rho(k)}{\sum_{k=1}^{K} \rho(k)}\right] \tag{3.135}$$

where $\rho(k)$ are the weights. We shall simplify the calculation by employing weights, the sum of which is independent of W. Sakoe and Chiba[61] have suggested the use of

$$\rho(k) = i(k) - i(k-1) + j(k) - j(k-1) \tag{3.136A}$$

which yields

$$\sum_{k=1}^{K} \rho(k) = I + J \tag{3.136B}$$

The weights are shown in Figure 17. The performance index (Equation 3.135) becomes:

$$D(x(t_i),x(t_j)) = \frac{1}{I+J}\underset{W}{\mathrm{Min}}\left(\sum_{k=1}^{K} d(c(k))\rho(k)\right) \tag{3.137}$$

The dynamic programming procedure allows us to perform the minimization in an efficient manner. Define the measure, $g(i(m),j(m))$:

$$g(i(m),j(m)) = \mathrm{Min}\left(\sum_{k=1}^{m} d(c(k))\rho(k)\right) \tag{3.138A}$$

The dynamic programming[65] (DP) equation proves the $g(i(m),j(m))$ measure by means of $g(i(m-1),j(m-1))$:

$$g(i(m),j(m)) = \underset{c(m-1)}{\mathrm{Min}}\left(g(i(m-1),j(m-1)) + d(c(m))\rho(m)\right) \tag{3.138B}$$

The point, $c(m)$, on the warping function will be determined by considering all allowable routes into the point $c(m)$. The route that minimizes $g(i(m),j(m))$ is chosen. It is clear that the more constraints we pose on the warping function the less routes we shall have to check.

The procedure starts with the initial conditions:

$$g(c(1)) = d(c(1)) \cdot \rho(1) \tag{3.139A}$$

and finally the distance is

$$D(x(t_i),x(t_j)) = \frac{1}{I+J} g_K(c(K)) \tag{3.139B}$$

Hence, for the weights and constraints we have imposed, we get the algorithm initial conditions:

$$g(1,1) = 2d(1,1) \tag{3.140A}$$

the DP equation:

$$g(i,j) = \text{Min}\begin{cases} g(i,j-1) + d(i,j) \\ g(i-1,j-1) + 2d(i,j) \\ g(i-1,j) + d(i,j) \end{cases} \tag{3.140B}$$

and the search window:

$$\frac{I}{J}(j-\gamma) \leq i \leq \frac{I}{J}(j+\gamma) \tag{3.140C}$$

The distance between warped signals is

$$D(x(t_i)x(t_j)) = \frac{1}{I+J}(g(I,J)) \tag{3.140D}$$

Equations 3.140A to D are recurrently calculated in ascending order, scanning the search window in the order:

$$j = 1,2,\ldots,J$$
$$i = \text{Max}\left(1, \frac{I}{J}(j-\gamma)\right),\ldots,\text{Min}\left(I, \frac{I}{J}(j+\gamma)\right) \tag{3.141}$$

The search is conducted, row by row, beginning with $j = 1$ and $i = 1,2, \ldots, (1 + \gamma)I/J$, followed by $j = 2$, $i = 1,2, \ldots, (2 + \gamma)I/J$, thus scanning all of the search window. For each point scanned, g(i,j) is calculated, until the endpoint (I,J) is reached. The algorithm yields directly the distance measure between the time warped words.

Note that when calculating the measures of the *ith* row, previously calculated measures from the same row or from the row $(i - 1)$ only are needed. Hence, one has to store the measures of current and previous rows only. At the most, this means the storage of $2\gamma I/J$ numbers.

The procedure described above yields the distance measure between the two warped signals. It does not yield the warping function itself. This is so because the optimal route W(c(k)) can be evaluated only after the search window has been completely scanned. In the procedure described above we have not stored the measures g(i,j). The optimal route can thus not be retrived. Consider the following modification to the described algorithm. After calculating the measure g(i,j) by Equation 3.140B, we record and store the choice made by

$$w(i,j) = q$$

where

$$q = \begin{cases} 0 \ , \ g(c(k - 1)) = g(i,j - 1) \\ 1 \ , \ g(c(k - 1)) = g(i - 1,j - 1) \\ 2 \ , \ g(c(k - 1)) = g(i - 1,j) \end{cases} \tag{3.142}$$

the value of w(i,j) tells us from what allowable point we have reached the current point (i,j). Hence, each point in the search window has attached to it information about the optimal route to reach it. After scanning has terminated, we can reconstruct the optimal route W(c(k)) by going backward from w(I,J) to w(1,1). For example, if w(c(K)) = w(I,J) = 2, we know that c(K − 1) = (I − 1,J). To proceed we check w(I − 1,J) and so on.

The storage requirement for this procedure is much higher, of course. For each point in the search window we require a storage of 0.25 bytes (q requires only two bits). The total number of points in the search window is $\gamma I(2 - \gamma/J)$ (for the case $J > \gamma/2$) and required storage is thus $0.25\ \gamma I(2 - \gamma/J)$ bytes.

Time warping by means of ordered graph search (OGS) technique has been developed.[63] It has been shown that the OGS algorithm can solve the time warping problem with essentially the same accuracy as the DP algorithm, with computations reduced by a factor of about 2.5. This reduction in computation, however, is attained at the expense of a more complicated combinatoric effort. It has been argued[63] therefore that when special high-speed hardware is used for the computation, the OGS may have no advantage over the DP.

REFERENCES

1. **Duda, P. O. and Hart, P. E.,** *Pattern Classification and Scene Analysis,* Wiley-Interscience, New York, 1973.
2. **Young, T. Y. and Calvert, T. W.,** *Classification, Estimation and Recognition,* Elsevier, New York, 1974.
3. **Fukunaga, K.,** *Introduction to Statistical Pattern Recognition,* Academic Press, New York, 1972.
4. **Tou, J. T. and Gonzalez, R. C.,** *Pattern Recognition Principles,* Addison-Wesley, Reading, Mass., 1974.
5. **Ahmed, N. and Rao, K. R.,** *Orthogonal Transforms for Digital Signal Processing,* Springer-Verlag, Berlin, 1975.
6. **Fu, K. S., Ed.,** *Applications of Pattern Recognition,* CRC Press, Boca Raton, Fla., 1982.
7. **Chen, C. H., Ed.,** *Digital Waveform Processing and Recognition,* CRC Press, Boca Raton, Fla., 1982.
8. **Fu, K. S., Ed.,** Special issue on feature extraction and selection in pattern recognition, *IEEE Trans. Comput.,* 20, 1971.
9. **Okada, M. and Maruyama, N.,** Software system for real time discrimination of multi-unit nerve impulses, *Comput. Prog. Biomed.,* 14, 157, 1981.
10. **Gevins, A. S.,** Pattern recognition of human brain electrical potential, *IEEE Trans. Pattern Anal. Mach. Intelligence,* 2, 383, 1980.
11. **Gersch, W., Yonemoto, J., and Naitoh, P.,** Automatic classification of multivariate EEG's using an amount of information measure and the eigenvalues of parametric time series model features, *Comput. Biomed. Res.,* 10, 297, 1977.
12. **Bodenstein, G. and Schneider, W.,** Pattern recognition of clinical electroencephalograms, in *Proc. Int. Conf. Signal Process.,* Florence, 1981, 206.
13. **Ferber, G.,** An adaptive system for the automatic description of EEG background activity, *Method. Inf. Med.,* 20, 32, 1981.
14. **Childers, D. G., Bloom, P. A., Arroyo, A. A., Roucos, S. E., Fischler, I. S., Achariyapaopan, T., and Perry, N. W.,** Classification of cortical responses using features from single EEG records, *IEEE Trans. Biomed. Eng.,* 29, 423, 1982.
15. **Yunk, T. P. and Tuteur, F. B.,** Comparison of decision rules for automatic EEG classification, *IEEE Trans. Pattern Anal. Mach. Intelligence,* 2, 420, 1980.
16. **Lim, A. J. and Winters, W. D.,** A practical method for automatic real-time EEG sleep state analysis, *IEEE Trans. Biomed. Eng.,* 27, 212, 1980.

17. **Lam, C. F., Zimmermann, K., Simpson, R. K., Katz, S., and Blackburn, J. G.,** Classification of somatic evoked potentials through maximum entropy spectral analysis, *Electroencephalogr. Clin. Neurophysiol.*, 53, 491, 1982.
18. **Gersch, W., Martinelli, F., and Yonemoto, J.,** Automatic classification of EEG, Kullback-Leibler nearest neighbor rules, *Science*, 205(4402), 193, 1979.
19. **Ruttimann, U. E.,** Compression of the ECG by prediction or interpolation and entropy encoding, *IEEE Trans. Biol. Med. Eng.*, 26(11), 613, 1979.
20. **Metaxaki-Kossionides, C., Athenaios, S. S., and Caroubalos, C. A.,** A method for compression reconstruction of ECG signals, *J. Biomed. Eng.*, 3, 214, 1981.
21. **Abenstein, J. P. and Tompleins, W. J.,** A new data reduction algorithm for real time ECG analysis, *IEEE Trans. Biomed. Eng.*, 29, 43, 1982.
22. **Jain, U., Rautaharju, P. M., and Warren, J.,** Selection of optimal features for classification of electrocardiograms, *J. Electrocardiogr.*, 14, 239, 1981.
23. **Kuklinski, W. S.,** Fast Walsh transform data — compression algorithm: ECG applications, *Med. Biol. Eng. Comput.*, 21, 465, 1983.
24. **Nygards, M. E. and Hulting, J.,** An automated system for ECG monitoring, *Comput. Biomed. Res.*, 12, 181, 1979.
25. **Shridar, M. and Stevens, M. F.,** Analysis of ECG data for data compression, *Int. J. Biomed. Comput.*, 10, 113, 1979.
26. **Pahlm, O., Borjesson, P. O., and Werner, O.,** Compact digital storage of ECG's, *Comput. Prog. Biomed.*, 9, 293, 1979.
27. **Cashman, P. M. M.,** A pattern recognition program for continuous ECG processing in accelerated time, *Comput. Biomed. Res.*, 11, 311, 1978.
28. **Gustafson, D. E., Willsky, A. S., Wang, J. Y., Lancester, M. C., and Triebwasser, J. H.,** ECG/VCG rhythm diagnosis using statistical signal analysis. I. Identification of persistent rhythms. II. Identification of transient rhythms, *IEEE Trans. Biomed. Eng.*, 25, 344, 1978.
29. **Womble, M. E., Halliday, J. S., Mitter, S. K., Lancester, M. C., and Triebwasser, J. H.,** Data compression for storing and transmitting ECG's/VCG's, *Proc. IEEE*, 65, 702, 1977.
30. **Ahmed, N., Milne, P. J., and Harris, S. G.,** Electrocardiographic data compression via orthogonal transforms, *IEEE Trans. Biomed. Eng.*, 22, 484, 1975.
31. **Young, T. Y. and Huggins, W. H.,** Computer analysis of electrocardiograms using a linear regression technique, *IEEE Trans. Biomed. Eng.*, 21, 60, 1964.
32. **Marcus, M., Hammerman, H., and Inbar, G. F.,** ECG classification by signal expansion on orthogonal K-L bases, Paper 9.25, in Proc. IEEE *Melecon* Conf., Tel-Aviv, Israel, May 1981.
33. **Iwata, A., Suzumura, N., and Ikegaja, K.,** Pattern classification of the phonocardiograms using linear prediction analysis, *Med. Biol. Eng. Comput.*, 15, 407, 1977.
34. **Urquhart, R. B., McGhee, J., Macleod, J. E. S., Banham, S. W., and Moran, F.,** The diagnostic value of pulmonary sounds: a preliminary study by computer aided analysis, *Comput. Biol. Med.*, 11, 129, 1981.
35. **Cohen, A. and Landsberg, D.,** Analysis and automatic classification of breath sounds, *IEEE Trans. Biomed. Eng.*, 31, 585, 1984.
36. **Inbar, G. F. and Noujaim, A. E.,** On surface EMG spectral characterization and its application to diagnostic classification, *IEEE Trans. Biomed. Eng.*, 31, 597, 1984.
37. **Childers, D. G.,** Laryngeal pathology detection, *CRC Crit. Rev. Bioeng.*, 2, 375, 1977.
38. **Cohen, A. and Zmora, E.,** Automatic classification of infants' hunger and pain cry, in *Proc. Int. Conf. Digital Signal Process.*, Cappellini, V. and Constantinides, A. G., Eds., Elsevier, Amsterdam, 1984.
39. **Annon, J. I. and McGillen, C. G.,** On the classification of single evoked potential using a quadratic classifier, *Comput. Prog. Biomed.*, 14, 29, 1982.
40. **Fukunaga, K. and Koontz, W. L. G.,** Application of the Karhunen-Loeve expansion, *IEEE Trans. Comput.*, 19, 311, 1970.
41. **Mausher, M. J. and Landgrebe, D. A.,** The K-L expansion as an effective feature ordering technique for limited training sample size, *IEEE Trans. Geosci. Rem. Sens.*, 21, 438, 1983.
42. **Fernando, K. V. M. and Nicholson, H.,** Discrete double sided K-L expansion, *IEE Proc.*, 127, 155, 1980.
43. **Bromm, B. and Scharein, E.,** Principal component analysis of pain related cerebral potentials to mechanical and electrical stimulation in man, *Electroencephalogr. Clin. Neurophysiol.*, 53, 94, 1982.
44. **Oja, E. and Karhunen, J.,** Recursive construction of Karhunen-Loeve expansions for pattern recognition purposes, in *Proc. IEEE Pattern Recog. Conf.*, Miami, 1980, 1215.
45. **Klemma, V. C. and Laub, A. J.,** The SVD, its computation and some applications, *IEEE Trans. Autom. Control*, 25, 164, 1980.
46. **Tou, J. T. and Heydorn, R. P.,** Some approaches to optimum feature extraction, in *Computers and Information Sciences*, Vol. 2, Tou, J. T., Ed., Academic Press, New York, 1967.

47. **Haimi-Cohen, R. and Cohen, A.**, A microcomputer controlled system for stimulation and acquisition of evoked potentials, *Comput. Biomed. Res.*, in press.
48. **Tufts, R. W., Kumaresan, R., and Kirsteins, I.**, Data adaptive signal estimation by SVD of data matrix, *Proc. IEEE*, 70, 684, 1982.
49. **Tominaga, S.**, Analysis of experimental curves using SVD, *IEEE Trans. Acoust. Speech Signal Process.*, 29, 429, 1981.
50. **Shlien, S.**, A method for computing the partial SVD, *IEEE Trans. Pattern Anal. Mach. Intelligence*, 4, 671, 1982.
51. **Damen, A. A. and van der Kam, J.**, The use of the SVD in electrocardiography, *Med. Biol. Eng. Comput.*, 20, 473, 1982.
52. **Foley, D. H. and Sammon, J. W.**, An optimal set of discriminant vectors, *IEEE Trans. Comput.*, 24, 281, 1975.
53. **Cox, J. R., Nolle, F. M., and Arthur, R. M.**, Digital analysis of the EEG, the blood pressure wave and the ECG, *Proc. IEEE*, 60, 1137, 1972.
54. **Noble, B. and Daniel, J. W.**, *Applied Linear Algebra*, 2nd ed., Prentice-Hall, Englewood Cliffs, N.J., 1977.
55. **Haimi-Cohen, R. and Cohen, A.**, On the computation of partial SVD, in *Proc. Int. Conf. Digital Sig. Proc.*, Cappellini, V. and Constantinides, A. G., Eds., Elsevier, Amsterdam, 1984.
56. **Cheung, R. S. and Eisenstein, B. A.**, Feature selection via dynamic programming for text-independent speaker identification, *IEEE Trans. Acoust. Speech. Signal Process.*, 26, 397, 1978.
57. **Chang, C. Y.**, Dynamic programming as applied to feature subset selection in pattern recognition system, *IEEE Trans. Syst. Man Cybern.*, 3, 166, 1973.
58. **Shrdhar, M., Baraniecki, M., and Mohanlerishman, N.**, A unified approach to speaker verification, *Speech Commun.*, 1, 103, 1982.
59. **Cohen, A. and Froind, T.**, Software package for interactive text-independent speaker verification, Paper 6.2.3, in *Proc. IEEE MELECON Conf.*, Tel-Aviv, Israel, 1981.
60. **Sakoe, H. and Chiba, S.**, Dynamic programming algorithm optimization for spoken word recognition, *IEEE Trans. Acoust. Speech Signal Process.*, 26, 43, 1978.
61. **Sakoe, H.**, Two level DP matching — a dynamic programming based pattern matching algorithm for connected work recognition, *IEEE Trans. Acoust. Speech Signal Process.*, 27, 588, 1979.
62. **Paliwal, K. K., Agarwal, A., and Sinha, S. S.**, A modification over Sakoe and Chiba's dynamic time warping algorithm for isolated word recognition, *Signal Process.*, 4, 329, 1982.
63. **Brown, M. K. and Rabiner, L. R.**, An adaptive, ordered, graph search technique for dynamic time warping for isolated word recognition, *IEEE Trans. Acoust. Speech Signal Process.*, 30, 535, 1982.
64. **Rabiner, L. R., Rosenberg, A. E., and Levinson, S. E.**, Considerations in dynamic time warping algorithms for discrete word recognition, *IEEE Trans. Acoust. Speech Signal Process.*, 26, 575, 1978.
65. **Bellman, R. and Dreyfus, S., Eds.**, *Applied Dynamic Programming*, Princeton University Press, Princeton, N.J., 1962.

Chapter 4

SYNTACTIC METHODS

I. INTRODUCTION

Two general approaches are known for the problem of pattern (and signal) recognition. The first, and better known one, is the decision-theoretic, or discriminant, approach (see Chapter 3) and the second is the syntactic, or structural, approach.

In the first approach the signal is represented by a set of features describing the characteristics of the signal which are of particular interest. For example, when analyzing the speech signal for the purpose of detecting laryngial disorders, features must be defined and extracted which are independent of the text (as much as possible) and are dependent on the anatomy of the physiological system under test. The features set serves to compress the data and reduce redundancy. This general method with its biomedical applications is discussed in Chapter 3.

The syntactic method[1-3] uses structural information to define and classify patterns. Syntactic methods have been applied to the general problem of scene analysis, e.g., to the automatic recognition of chromosomes and finger prints. It has also been applied to signal analysis,[4-9] with applications to such areas as seismology[10] and biomedical signal processing. Syntactic methods have been applied to EEG analysis,[11-14] ECG analysis,[15-22,29] the classification of the carotid waveform,[23] and to speech processing.[24]

The syntactic approach has gained a lot of attention from researchers in the field of scene analysis since this approach possesses the structure-handling capability which seems to be important in analyzing patterns and scenes. For exactly the same reasons, this approach seems to have a good potential in analyzing complex biological signals. The human interpreter of biological signals, the electrocardiographer or electroencephalographer, e.g., when analyzing the signals, observes the structure of the waveforms for his diagnostic decision. The human diagnosis process is thus more closely related to the syntactic approach than to the more conventional decision-theoretic approach.

The syntactic approach is also known by the terms linguistic, structural, and grammatical approach. An analog between the structure of a pattern and the syntax of a language can be drawn. A pattern is described by the relationships between simple subpatterns from which it is composed. The rules describing the composition of the subpatterns are expressed in a similar manner to grammar rules in linguistics.

The basic ideas behind the syntactic approach are similar in principal to those of the decision-theoretic approach. In order to classify a signal into several known classes, the structure of each class must be learned. In a supervised learning mode, known samples of signals from each class are provided such that the structural features (primitives) and the rules of their combination into the given signal (grammar) can be estimated. These are stored in the system. An unknown signal, to be analyzed and classified, is subjected to some preprocessing in order to reduce noise, and its primitives are extracted. Classification is made by applying the syntax of each one of the classes. By means of some measure, a decision is made as to the best syntax that fits the signal. Figure 1 shows schematically a general syntactic signal recognition system.

Example 4.1

As an example consider the signal in Figure 2a with the primitives (features) defined in Figure 2b. The signal can be described by a string of primitives in the example: aabbccdeffgggfcccccaacbbcccc. The string representation may be sufficient to describe simple waveforms. More complex waveforms are described by means of an hierarchical

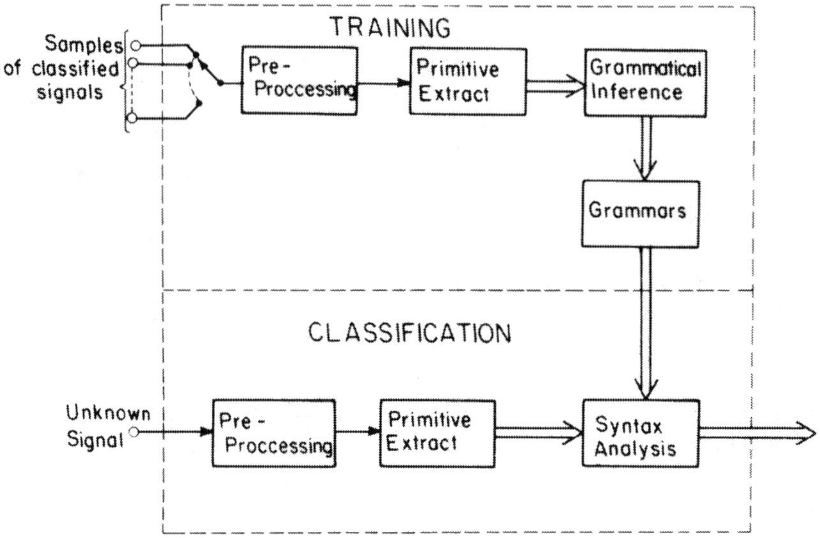

FIGURE 1. A general syntactic recognition system.

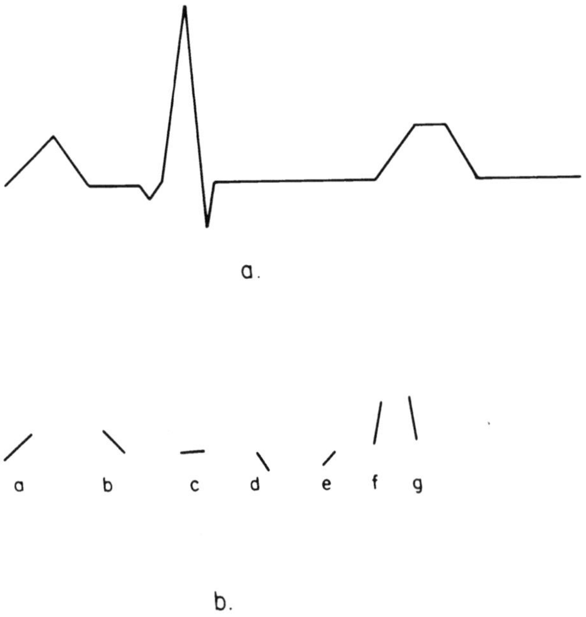

FIGURE 2. Syntactic representation of ECG. (a) Piecewise linear model of ECG; (b) primitives.

structural tree. Consider the signal in Figure 3a. It has been segmented into six sections. Each section is to be described by the primitives presented in Figure 3b and the complete waveform described by the structural tree of Figure 3c. Another possible set of primitives for this example is shown in Figure 3d.

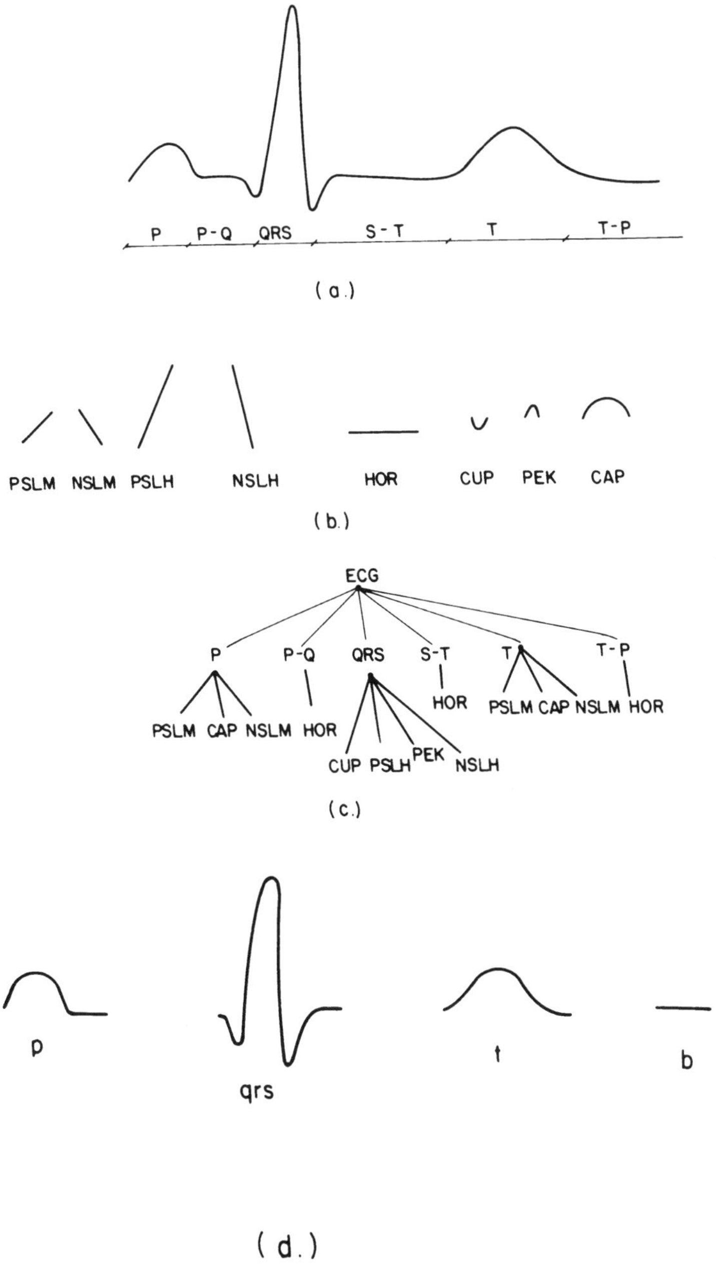

FIGURE 3. Syntactic representation of ECG. (a) ECG model; (b) a set of eight primitives; (c) relational tree; (d) an alternative set of primitives.

II. BASIC DEFINITIONS OF FORMAL LANGUAGES

The description of the structure of the signal is performed by grammars or syntax rules. Grammar can also be used to describe all the signals (sentences) belonging to a given class (language). Usually a class is represented by a given set of known signals (training set). It is required to estimate the class generating grammar from the training set — a process known as grammatical inference.

We shall denote the grammar, G, as the grammar that can model the source generating the signal set. All the sentences that can be generated by the grammar G constitute the set L(G) — the language generated by G. A phrase structure grammar,[1] G, is a quadruple given by:

$$G = (V_N, V_T, R, \sigma) \tag{4.1}$$

where:

V_N = finite set of nonterminal symbols (vocabularies or variables) of G;
V_T = finite set of terminal symbols;
σ = start symbol ($\sigma \in V_N$); and
R = finite set of rewriting rules, or productions, denoted by $\alpha \to \beta$, where α and β are strings.

Several notations have been introduced for syntactic analysis:

1. $\alpha \to \beta$: the symbol $\to$ means "can be rewritten as" and is known as production.
2. x^n: if x is a string, x^n is x written n times.
3. $|x|$: is the number of symbols in the string x, denoted also as the length of the string.
4. V_T^*: a set of all finite length strings which can be generated using the string V_T.

Phrase structure grammars were divided into four types:[1]

1. Unrestricted grammars (type 0) are grammars with no restrictions placed on the productions. Type 0 grammars are too general and have not been applied much.
2. Context-sensitive grammars (type 1) are grammars in which the productions are restricted to the form:

$$\xi_1 A \xi_2 \to \xi_1 \beta \xi_2 \tag{4.2}$$

where $A \in V_N$ and $\xi_1, \xi_2, \beta \in V^*$. The languages generated by type 1 grammars are called context sensitive languages.
3. Context-free grammars (type 2) are grammars in which the productions are restricted to the form:

$$A \to \beta \tag{4.3}$$

where $A \in V_N$ and $\beta \in V^*$. Here β is replacing A independently of the context in which A appears. Languages generated by type 2 grammars are called context-free languages.
4. Finite state, or regular grammar (type 3) are grammars in which the productions are restricted to $A \to aB$ or $A \to b$ where $A,B \in V_N$ and $a,b \in V_T$ (all are single symbols and not strings).

Example 4.2
Consider, for example, the finite state grammar $G_{F1} = \{V_{N2}, V_T, R_2, \sigma\}$, where

$$V_{N2} = \{\sigma\}$$

$$V_T = \{a,b\}$$

$$\text{and } R_2: \sigma \to a\sigma$$

$$\sigma \to b$$

The finite state language generated by grammar G_{F1} is $L(G_{F1}) = \{a^n b | n = 1,2, \ldots\}$.

Example 4.3

Consider the grammar $G_{C1} = (V_{N3}, V_T, R_3, \sigma)$ where $V_{N3} = \{\sigma, A\}$, $V_T = \{a,b\}$, and

$$R_3: \sigma \rightarrow Ab$$

$$A \rightarrow Aa$$

$$A \rightarrow a$$

This is a context-free grammar, since it obeys Equation 4.3. The language generated by it is the language:

$$L(G_{C1}) = \{a^n b | n = 1,2,\ldots\} \quad (4.4)$$

which is the language consisting of strings with n "a's" followed by one "b". Note that this language is the same as the finite state language $L(G_{F1})$ of the previous example. Different grammars can generate the same language.

Example 4.4

Consider another example[1] with $G_{C2} = (V_{N4}, V_T, R_4, \sigma)$, where $V_{N4} = \{\sigma, A, B\}$, $V_T = \{a,b\}$, and

R_4: (1) $\sigma \rightarrow aB$	(5) $A \rightarrow a$
(2) $\sigma \rightarrow bA$	(6) $B \rightarrow b$
(3) $A \rightarrow a\sigma$	(7) $B \rightarrow aBB$
(4) $A \rightarrow bAA$	(8) $B \rightarrow b$

Grammar G_{C2} is context free since it obeys Equation 4.3, namely, each production in R_4 has a nonterminal to its left and to its right a string of terminal and nonterminal symbols. Examples of sentences generated by G_{C2} are $\{(ab)^n\}$ by activation $(n - 1)$ times rules 1 and 7 followed by 8, or {ba} by activating rules 2 and 5. In general, the language $L(G_{C2})$ is the set of all words with an equal number of a's and b's.

An alternate method to describe a finite state grammar is by a graph known also as the state transition diagram. The graph consists of nodes and paths. The nodes correspond to states, namely, the nonterminal symbols of V_N, and a special node T (the terminal node). Paths exist between nodes N_i and N_j for every production if R of the type $N_i \rightarrow aN_j$. Paths to the terminal node T from node A_i exist for each production $A_i \rightarrow a$.

Example 4.5

Consider the finite state grammar $G_{F2} = \{V_{N5}, V_T, R_5, \sigma\}$ with $V_{N5} = \{\sigma, A, B\}$, $V_T = \{a,b\}$, and

R_5: (1) $\sigma \rightarrow aA$	; (5) $A \rightarrow a$
(2) $\sigma \rightarrow b$	; (6) $B \rightarrow aB$
(3) $A \rightarrow bA$	; (7) $B \rightarrow b$
(4) $A \rightarrow aB$	;

The graph of G_{F2} is shown in Figure 4.

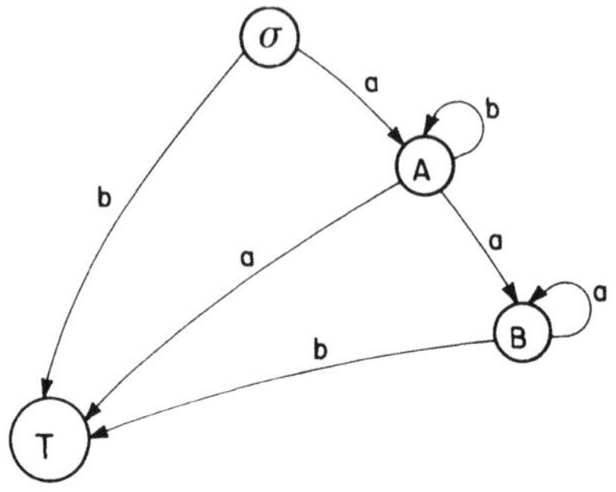

FIGURE 4. Finite state grammar, G_{F2}.

III. SYNTACTIC RECOGNIZERS

A. Introduction

The signals under testing are represented by strings that were generated by a grammar. Each class has its own grammar. It is the task of the recognizer to determine which of the grammars has produced the given unknown string. Consider the case where M classes are given, w_i, i = 1,2, . . . ,M, each with its grammar, G_i. The process known as syntax analysis, or parsing, is the process that decides whether the unclassified string x belongs to the language $L(G_i)$, i = 1,2, . . . ,M. If it has been determined that $x \in L(G_j)$, then x is classified into w_j.

We shall consider first the recognition of strings by automata. Recognizing automata have been developed for the various types of phrase structure grammars. Of interest here are the finite automaton, used to recognize finite state grammars and the push-down automaton used to recognize context-free grammars. The discussion of more general parsing methods will follow.

B. Finite State Automata

A deterministic finite state automata, A, is a quintuple,

$$A = (\Sigma, Q, \delta, q_0, F) \tag{4.5}$$

where Σ is the alphabet — a final set of input symbols, Q is a final set of states, δ is the mapping operator, q_0 is the start state, and F is a set of final states.

The automaton operation can be envisioned as a device, reading data from a tape. The device is initially at state q_0. The sequence, x, written on the tape is read, symbol by symbol, by the device. The device moves to another state by the mapping operator:

$$\delta(q_1, \xi) = q_2$$

$$\xi \in \Sigma \tag{4.6A}$$

which is interpreted as: the automaton is in state q_1 and upon reading the symbol ξ moves to state q_2. The string x is said to be accepted by the automaton A, if upon reading the complete string x, the automaton is in one of the final states.

The transformation operation (Equation 4.6) can be extended to include strings. The string x will thus be accepted or recognized by automaton A, if:

$$\delta(q_0, x) = p \text{ for some } p \in F \tag{4.6B}$$

namely, starting from state q_0 and scanning the complete string x, the automaton A will follow a sequence of states and will halt at state p which is one of the final states.

Example 4.6

Consider a deterministic finite state automaton, A_1, given by

$$A_1 = (\Sigma_1, Q_1, \delta, q_0, F_1), \text{ with } \Sigma_1 = \{a,b\}$$

$$Q_1 = \{q_0,q_1,q_2,q_3,q_4\} \text{ and } F_1 = \{q_3,q_4\}$$

The state transition mapping of A_1 is

$$\delta(q_0,a) = \{q_1\} \qquad \delta(q_1,b) = \{q_0\}$$

$$\delta(q_0,b) = \{q_3\} \qquad \delta(q_2,a) = \{q_3\}$$

$$\delta(q_1,a) = \{q_2\} \qquad \delta(q_2,b) = \{q_4\}$$

The state transition diagram of A_1 is given in Figure 5. Note that terminal states are denoted by two concentric circles. The strings $x_1 = \{(ab)^n b\}$, $x_2 = \{(ab)^m a^2\}$ are recognized by A_1 since $\delta(q_0,x_i) = \{q_3\} \in F$, i = 1,2; the string $x_3 = \{(ab)^n ab\}$ is recognized since $\delta(q_0,x_3) = \{q_4\} \in F$.

A nondeterministic finite state automaton is the same as the deterministic one, except for the fact that the transformation $\delta(q,\xi)$ is a set of states rather than a single state, as indicated in Equation 4.6. Hence, for the nondeterministic automaton:

$$\delta(q,\xi) = \{q_1,q_2,\ldots,q_n\} \tag{4.7}$$

Equation 4.7 describes the transformation of a nondeterministic automaton at state q upon reading the symbol ξ. The automaton will move into any one of the states $q_1,q_2, \ldots ,q_n$. It is assumed that the automaton always chooses the correct state to jump to.

Example 4.7

Consider the nondeterministic finite state automaton $A_2 = (\Sigma_1,Q_2,\delta,q_0,F_2)$ with $\Sigma_1 = \{a,b\}$, $Q_2 = \{q_0,q_1,q_2,q_3\}$, and $F_2 = \{q_3\}$. The state transition mapping of A_2 is given by the transformations:

$$\delta(q_0,a) = \{q_1\} \qquad \delta(q_1,b) = \{q_1\}$$

$$\delta(q_0,b) = \{q_3\} \qquad \delta(q_2,a) = \{q_2\}$$

$$\delta(q_1,a) = \{q_2,q_3\} \qquad \delta(q_2,b) = \{q_3\}$$

The state transition diagram of A_2 is given in Figure 6. Note that the transformation $\delta(q_1,a) = \{q_2,q_3\}$ makes the automaton a nondeterministic one.

The strings $x_4 = \{abaab\}$, $x_5 = \{ab^m a^n b\}$ will be recognized by A_2, since $\delta(q_0,x_i) = \{q_3\}$, i = 4,5. Note that the automaton A_2 will recognize all strings generated by the grammar G_{F2} given in Figure 4. It can be shown[1] that for any nondeterministic finite state automaton, accepting a set of strings, L, there exists a deterministic automaton recognizing the same set of strings. It can also be shown that for any finite state grammar, G, there exists a finite state automaton that recognizes the language generated by G.

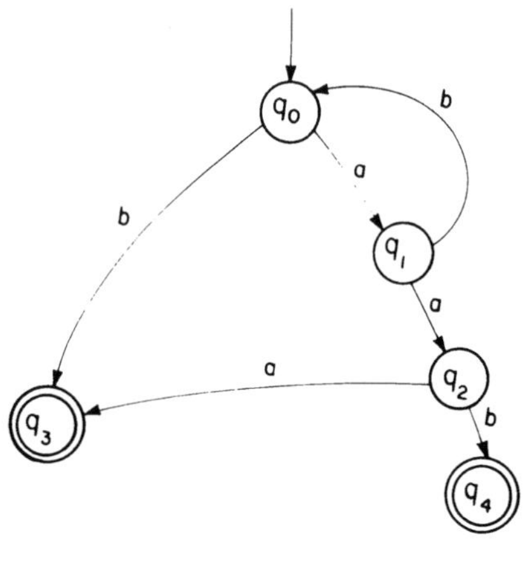

FIGURE 5. State transition diagram of finite state automaton, A.

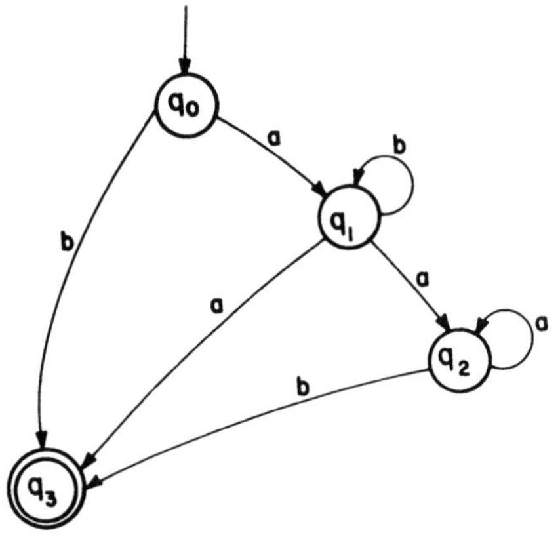

FIGURE 6. State transition diagram of nondeterministic finite state automaton, A_2.

Example 4.8

Consider now the ECG signal shown in Figure 3A with the primitives of Figure 3D (the example is based on an example given by Gonzalez and Thompson[2]). A regular grammar describing the normal ECG is given by

$$G = (\{\sigma, A, B, C, D, E, F, G, H\}, \{p, qrs, t, b\}, R, \sigma)$$

$$R: \sigma \rightarrow pA \quad A \rightarrow qrs\ C \quad A \rightarrow bB$$

$$B \rightarrow qrs\ C \quad C \rightarrow bD \quad D \rightarrow tF$$

$$D \rightarrow bE \quad E \rightarrow tF \quad F \rightarrow b$$

$$F \rightarrow bG \quad G \rightarrow b \quad G \rightarrow bH$$

$$G \rightarrow pA \quad H \rightarrow b \quad H \rightarrow b\sigma$$

$$H \rightarrow pA$$

The normal ECG is defined here as the one having a basic complex (p qrs b t b) with normal variations including one b between the p and qrs waves, an additional b between the qrs

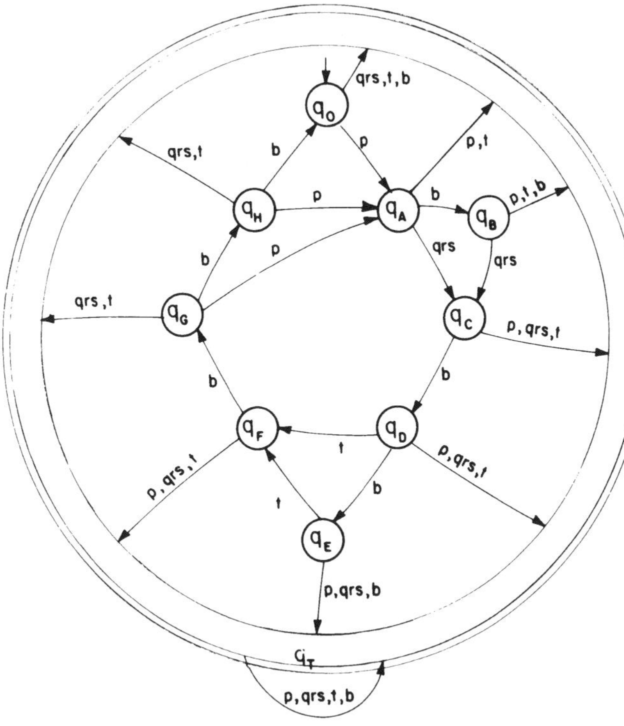

FIGURE 7. A deterministic finite state automaton for ECG analysis.

and t waves, and additional one or two b's between the t and the next p waves. A deterministic finite state automaton that recognizes normal ECG is shown in Figure 7. In this diagram a state q_T has been added to denote the terminal state.

C. Context-Free Push-Down Automata (PDA)

Context-free languages (that are not finite state), defined by Equation 4.3, cannot be recognized by the finite state automaton. One recognizer for context-free languages is the push-down automaton (PDA). The push-down automaton is similar to the finite state one, with the addition of a push-down stack. The stack is a "first in-last out" string storage. Strings can be stored in the push-down stack such that the first symbol is at the top. The automaton always reads the top symbol. Figure 8 shows a schematic diagram of the PDA. The stack is assumed to have infinite capacity.

The nondeterministic push-down automaton M is a septuple

$$M = (\Sigma,Q,\Gamma,\delta,q_0,Z_0,F) \tag{4.8}$$

with Σ, Q, q_0, and F the same as in Equation 4.5 and with Γ a finite set of push-down symbols; $Z_0 \in \Gamma$, a start symbol initially appearing on the push-down storage. The operator $\delta(q,\xi,Z)$ is a mapping operator:

$$\delta(q,\xi,Z) = \{(q_1,\gamma_1),(q_2,\gamma_2),\ldots,(q_m,\gamma_m)\} \tag{4.9}$$

where $q,q_1,q_2, \ldots ,q_m \in Q$ are states, $\xi \in \Sigma$ is an input symbol, Z is the current symbol at the top of the stack, and $\gamma_1,\gamma_2, \ldots ,\gamma_m \in \Gamma$ are strings of push-down symbols. The transformation (Equation 4.9) is interpreted as follows: the control is in state q, with the symbol

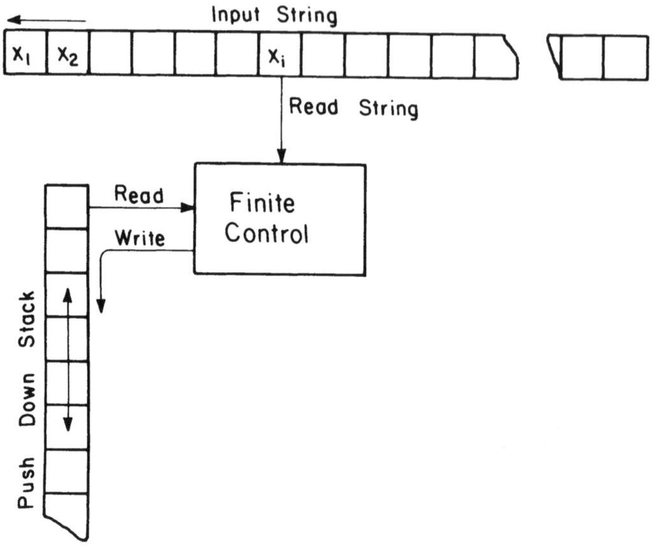

FIGURE 8. Push-down automaton.

Z at the top of the push-down stack, and the input symbol ξ is read from the input string. The control will choose one of the pairs (q_i,γ_i), $i = 1,2, \ldots ,m$, say (q_j,γ_j). It will replace the symbol Z in the stack by the string γ_j, such that its leftmost symbol appears at the top of the stack, will move to state q_j, and will read the next symbol of the input string.

If $\xi = \lambda$ (the null symbol), then (independent of the input string) the automaton, in state q, will replace Z by γ_j in the stack and will remain in state q. When $\gamma_j = \lambda$ the upper symbol of the stack is cleared. If the automaton reaches a step $\delta(q_i,\xi,\xi)$, namely, the uppermost symbol of the stack matches, the input symbol, ξ, is popped from the stack exposing the next in line stack symbol, η, with the next transformation $\delta(q_i,\gamma,\eta)$. If the automaton reaches a combination of state, input and state symbols for which no transformation is defined, it halts and rejects the input.

Acceptance of strings by the push-down automaton can be expressed in two ways: (1) The automaton reads all symbols of the input string without being halted and after the final input symbol has been moved into a state q belonging to the final set F, and (2) the automaton reads all input symbols without being halted. The transformation taken after reading the last input sample moves the automaton into a state $q \in Q$ with empty stack, namely, $\gamma = \lambda$. This type of acceptance is called "acceptance by empty store". For this case it is convenient to define the set of final states as the null set $F = \emptyset$.

Example 4.9

Consider a signal with primitives {a,b,c,d} as shown in Figure 9. Consider a class of signals generated with these primitives such that the language describing the class is the nonregular (nonfinite state) context-free language $\{x|x = ab^ncd^n, n \geq 0\}$. The members of the class of signals are depicted in Figure 9. The context-free grammar that generates the class of signal is

$$G_{C3} = (V_{N9}, V_{T9}, R_9, \sigma) = (\{\sigma,A\},\{a,b,c,d\},R_9,\sigma)$$

$$R_9: \quad \sigma \rightarrow aA$$

$$A \rightarrow bAd$$

$$A \rightarrow c \qquad (4.10)$$

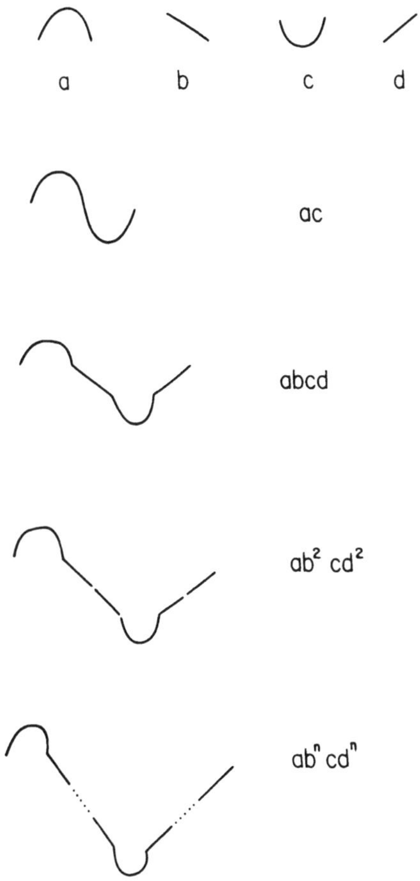

FIGURE 9. Primitives and samples of $L(G_{C3})$.

The push-down automaton that recognizes language $L(G_{C3})$ is the PDA M_1 given by

$$M_1 = (\Sigma,Q,\Gamma,\delta,q_0,Z_0,F) = (\{a,b,c,d\},\{q_0\},\{\sigma,A,B,C,D\},\delta,q_0,\sigma,\phi)$$

with the transition mapping:

$$(1)\ \delta(q_0,a,\sigma) = \{(q_0,DAB),(q_0,C)\}$$

$$(2)\ \delta(q_0,b,D) = \{(q_0,\lambda)\}$$

$$(3)\ \delta(q_0,b,A) = \{(q_0,AB),(q_0,CB)\}$$

$$(4)\ \delta(q_0,c,C) = \{(q_0,\lambda)\}$$

$$(5)\ \delta(q_0,d,B) = \{(q_0,\lambda)\}$$

$$(6)\ \delta(q_0,c,A) = \{(q_0,\lambda)\} \qquad (4.11)$$

Suppose that an input string $x_1 = \{abc\}$, that does not belong to G_{C3}, is checked by the automaton M_1. The following steps take place. Initially the stack holds the string σ; the input symbol read first is a; hence the first transformation is invoked. The automaton has to choose between replacing σ by DAB or by C (since this is a nondeterministic machine). If there is a right choice, it is assumed the automaton will take it.

Suppose the first choice is taken:

$$\delta(q_0,a,\sigma) = \{(q_0,DAB)\}$$

The automaton remains in state q_0; its stack now holds the string DAB with the symbol D in the uppermost (reading) location and the next input symbol (b) is read. The second transformation is invoked.

$$\delta(q_0,b,D) = \{(q_0,\lambda)\}$$

The automaton remains in state q_0; the symbol D is removed from the stack (replaced by the null symbol λ); the stack contains AB with the symbol A pushed into the uppermost stack location. The next input symbol, c, is read and the next transformation is

$$\delta(q_0,c,A) = \{(q_0,\lambda)\}$$

This transformation leaves the stack unempty, with symbol B. The input string has been all read. It, however, cannot yet be rejected since the first choice made may have been the wrong choice.

Return now to the first step and take the choice:

$$\delta(q_0,a,\sigma) = \{(q_0,C)\}$$

reading the next input symbol (b) and having C in the stack calls for transformation $\delta(q_0,b,C)$ which is undefined, thus causing the automaton to halt. The string {abc} is not recognized by the PDA, M_1.

Consider now the input string $x_2 = \{ab^2cd^2\}$ which does belong to the langauge $L(G_{C3})$. The following steps will be taken by M_1:

$$\delta(q_0,a,\sigma) = \{(q_0,DAB)\} \;;$$

$$\delta(q_0,b,D) = \{(q_0,\lambda)\} \quad ; \text{stack} = DAB$$

$$\delta(q_0,b,A) = \{(q_0,AB)\} \quad ; \text{stack} = AB$$

$$\delta(q_0,c,A) = \{(q_0,\lambda)\} \quad ; \text{stack} = ABB$$

$$\delta(q_0,d,B) = \{(q_0,\lambda)\} \quad ; \text{stack} = BB$$

$$\delta(q_0,d,B) = \{(q_0,\lambda)\} \quad ; \text{stack} = B$$

The last transformation replaces the symbol B in the stack by the null symbol λ and leaves it clear at the end of the input string. The string $x_2 = \{ab^2cd^2\}$ is therefore accepted by M_1.

In most applications of syntactic signal processsing, a class of signals is given (by means of their primitives and grammar) and a recognizer has to be designed to recognize the class of signals of interest. It has been proven[2] that for each context-free grammar, G, a PDA can be constructed to recognize L(G). The inverse statement, namely, for each PDA there is a grammar recognized by the automation, is also correct.

Consider the case where a context-free grammar, G, is given and it is desired to obtain the PDA that recognizes it. One relatively simple algorithm[2] is as follows: let $G = (V_N,V_T,R,\sigma)$ be the given context-free grammar. A PDA recognizing L(G) is $A = (V_T,\{q_0\},\Gamma,\delta,q_0,\sigma,\phi)$ with the push-down symbols Γ being the union of V_T and V_N and with the transformations δ obtained from the productions R by:

(1) If $A \rightarrow \alpha$ is in R,

then $\delta(q_0,\lambda,A) \rightarrow \{(q_0,\alpha)\}$,

where α is a string and λ is the null string.

(2) For each terminal a in V_T,

$$\delta(q_0,a,a) = \{(q_0,\lambda)\}. \tag{4.12}$$

Example 4.10

Consider the context-free grammar G_{C3} (Equation 4.10). The PDA, M_2, designed to recognize $L(G_{C3})$ by the rules (Equation 4.12), is

$$M_2 = (\{a,b,c,d\},\{q_0\},\{\sigma,a,b,c,d,A\},\delta,\ q_0,\sigma,\phi)$$

and the transformations are

$$\left.\begin{aligned} &\delta(q_0,\lambda,\sigma) \rightarrow \{(q_0,aA)\} \\ &\delta(q_0,\lambda,A) \rightarrow \{(q_0,bAd),(q_0,c)\} \end{aligned}\right\} \quad \text{By rule (1)}$$

$$\left.\begin{aligned} &\delta(q_0,a,a) \rightarrow \{(q_0,\lambda)\} \\ &\delta(q_0,b,b) \rightarrow \{(q_0,\lambda)\} \\ &\delta(q_0,c,c) \rightarrow \{(q_0,\lambda)\} \\ &\delta(q_0,d,d) \rightarrow \{(q_0,\lambda)\} \end{aligned}\right\} \quad \text{By rule (2)}$$

Consider again the input string $x_2 = \{ab^2cd^2\}$ belonging to $L(G_{C3})$. The PDA will proceed as follows:

$\delta(q_0,\lambda,\sigma) = \{(q_0,aA)\}$

$\delta(q_0,a,a) = \{(q_0,\lambda)\}$ — Stack = A

$\delta(q_0,\lambda,A) = \{(q_0,bAd),(q_0,c)\}$ — a is popped from stack

$\delta(q_0,\lambda,A) = \{(q_0,bAd)\}$ — Automaton selects (q_0,bAd), stack = bAd

$\delta(q_0,b,b) = \{(q_0,\lambda)\}$	b is popped from stack, stack = Ad
$\delta(q_0,\lambda,A) = \{(q_0,bAd)\}$	Automaton selects (q_0,bAd), d on stack is pushed down, stack = bAdd
$\delta(q_0,b,b) = \{(q_0,\lambda)\}$	b is popped from stack, stack = Add
$\delta(q_0,\lambda,A) = \{(q_0,c)\}$	stack = cdd
$\delta(q_0,c,c) = \{(q_0,\lambda)\}$	c is popped from stack, stack = dd
$\delta(q_0,d,d) = \{(q_0,\lambda)\}$	d is popped from stack, stack = d
$\delta(q_0,d,d) = \{(q_0,\lambda)\}$	Stack is empty, input string has been read — string accepted

D. Simple Syntax-Directed Translation

It is often necessary to map a string generated by one grammar into another string in another language. Such a mapping, based on underlying context-free grammars, is denoted a syntax-directed translation. With the help of simple syntax-directed translators, an input string may be mapped (classified) into an output string in addition to being recognized.

The simplest translator is the nondeterministic finite transducer given by the sextuple:

$$A_T = (\Sigma,Q,\Lambda,\delta,q_0,F) \tag{4.13}$$

where Σ, Q, q_0, and F are the same as in Equation 4.5, Λ is a finite output alphabet, and δ is the mapping operator. The translator can be seen as a finite state automaton with additional output tape onto which output mapping is written.

Example 4.11

The state diagram of a finite transducer for the problem of ECG analysis is given in Figure 10. This transducer not only recognizes normal ECGs (in the sense of Figure 7), but also classifies every ECG onto normal (N) and abnormal (Ab). In each step a symbol N or Ab is written on the output tape. For example, an input string pbqrsbbtbbb will be recognized by the transducer while placing the sequence N^9 in the output. An input string pbbqrsbbtbbb will produce the sequence N^2Ab^8 in the output. An input string that has been mapped into the output string with at least one Ab symbol is classified as abnormal.

Nonregular context-free syntax directed translation requires more sophisticated transducers. A nondeterministic push-down transducer is similar in structure to the finite transducer with a push-down stack added.

E. Parsing

General techniques for determining the sequence of productions used to derive a given string x, of context-free language L(G), exist. These techniques are called parsing or syntax

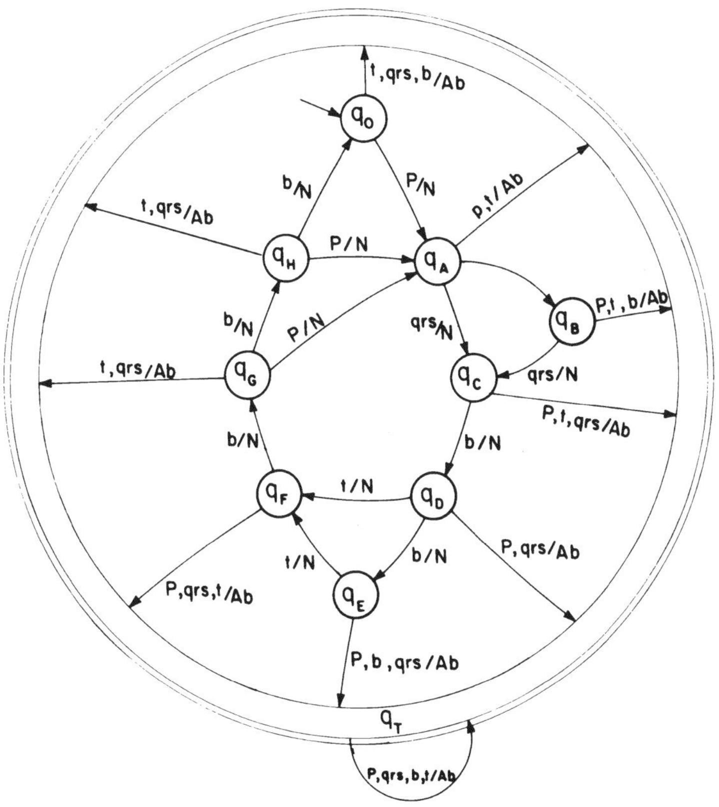

FIGURE 10. Finite transducer for ECG analysis.

analysis techniques. Two general approaches are known for parsing the string x. Bottom-up parsing starts from the string x and applies productions of G, in reverse fashion, in order to get to the starting symbol σ. Top-down parsing starts from the symbol σ and by applying the productions of G tries to get to the string x. Efficient parsing algorithms, such as the Cock-Yonger-Kasami algorithm, have been developed. The interested reader is referred to pattern recognition literature (e.g., References 1 and 2).

IV. STOCHASTIC LANGUAGES AND SYNTAX ANALYSIS

A. Introduction

Most often signal processing is done in a stochastic environment. Noise and uncertainties are introduced either due to the stochastic nature of the signal under test or due to the acquisition and primitive extraction processes. The classes of signals to be analyzed may overlap in the sense that a given signal may belong to several classes. Stochastic languages are used to solve such problems. If n grammars, G_i $i = 1,2, \ldots ,n$, are considered for the string x (representing the signal), the conditional probabilities $p(x|G_i)$, $i = 1,2, \ldots ,n$, are required. The grammar that most likely produced the string x is the grammar for which $p(G_j|x)$ is the maximum.

A stochastic grammar is one in which probability values are assigned to the various productions. The stochastic grammar G_s is a quintuple $G_s = \{V_N,V_T,R,P,\sigma\}$, where V_N, V_T, R, and σ are the same as in Equation 4.1 and P is a set of probabilities assigned to the productions of R.

We shall deal here only with unrestricted and proper stochastic grammars. An unrestricted stochastic grammar is one in which the probability assigned to a production does not depend on previous productions. Consider a nonterminal, T_i, for which there are m productions: $T_i \rightarrow \alpha_1, T_i \rightarrow \alpha_2, \ldots, T_i \rightarrow \alpha_m$; the productions are assigned probabilities, P_{ij}, j = 1,2, . . . ,m. A proper stochastic grammar is one in which

$$\sum_{j=1}^{m} P_{ij} = 1 \text{ for all i} \tag{4.14}$$

Stochastic grammars are divided into four types, in a similar manner to nonstochastic grammars (Equations 4.2 and 4.3). Therefore we speak of stochastic context-free and stochastic finite state grammars and languages.

Example 4.12

Consider the proper stochastic context-free grammar:

$$G_{s1} = (\{\sigma\},\{a,b\},R,\{p_1,(1 - p_1)\},\sigma)$$

$$R: \ (p_1)\ \sigma \rightarrow a\sigma a$$

$$(1 - p_1)\ \sigma \rightarrow bb$$

where the first production is assigned the probability p_1 and the second is assigned $(1 - p_1)$. The grammar is clearly a proper stochastic grammar. The grammar G_{s1} generates strings of the form $x_n = a^n bba^n$, $n \geqslant 1$. The probability of the string is $p(x_n) = p_1^n(1 - p_1)$.

B. Stochastic Recognizers

Finite state stochastic grammars can be recognized by a stochastic finite automaton. The automaton is defined by the sextuple:

$$A_s = (\Sigma,Q,\delta,q_0,F,P) \tag{4.15}$$

where Σ, Q, q_0, and F are the same as in Equation 4.5, P is a set of probabilities, and δ is the mapping operator to which probabilities are assigned. The stochastic finite automaton operates in a similar way to that of the finite one, except that the transition from one state to another is a random process with given probabilities. For an unrestricted automaton the probabilities do not depend on a previous state.

Example 4.13

Consider the finite state automaton of Figure 7, designed to recognize normal ECG. Assume that there is a probability of 0.1 that there will be no "t" wave present. We would still want to recognize this as a normal ECG. The automaton designed to recognize the signal is a modification of the finite state one. Its state transition diagram is shown in Figure 11. Note that another path has been added between states q_D and q_G and that each path is assigned an input symbol and a probability. When the automaton is in state q_D and the input symbol is "b", it can move (with probability of 0.9) to q_E or (with probability 0.1) to state q_G. The stochastic state transitions of the automaton are as follows:

$$\delta(q_0,p) = \{q_A\} \quad ; \quad p(q_A|p,q_0) = 1$$

$$\delta(q_A,qrs) = \{q_C\} \quad ; \quad p(q_c|qrs,q_A) = 1$$

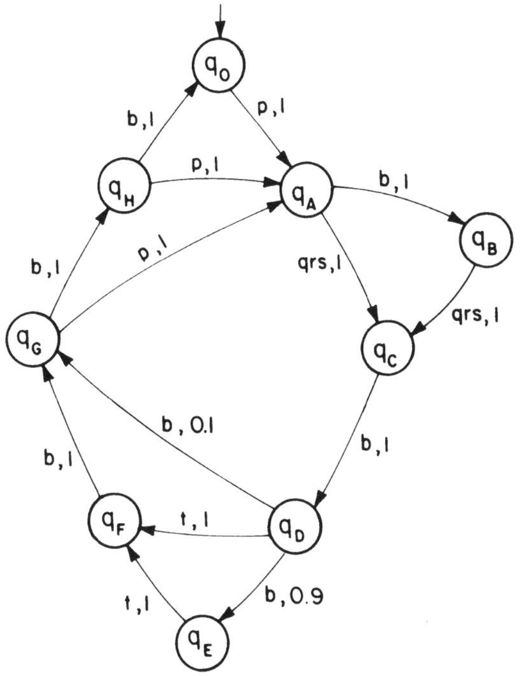

FIGURE 11. State transition diagram of stochastic finite state automaton for ECG analysis.

$$\delta(q_A,b) = \{q_B\} \quad ; \quad p(q_B|b,q_A) = 1$$

$$\delta(q_c,b) = \{q_D\} \quad ; \quad p(q_D|b,q_C) = 1$$

$$\delta(q_D,b) = \{q_E,q_G\} \quad ; \quad p(q_E|b,q_D) = 0.9$$

$$; \quad p(q_G|b,q_D) = 0.1$$

$$\delta(q_D,t) = \{q_F\} \quad ; \quad p(q_F|t,q_D) = 1$$

$$\delta(q_F,b) = \{q_G\} \quad ; \quad p(q_G|b,q_F) = 1$$

$$\delta(q_G,b) = \{q_H\} \quad ; \quad p(q_H|b,q_G) = 1$$

$$\delta(q_G,p) = \{q_A\} \quad ; \quad p(q_A|p,q_G) = 1$$

$$\delta(q_H,b) = \{q_0\} \quad ; \quad p(q_0|b,q_H) = 1$$

Stochastic push-down automata are designed to recognize stochastic context-free languages;

these are generalizations of the nondeterministic push-down automata. Simple stochastic syntax-directed languages can be recognized by stochastic syntax-directed translators. The interested reader is referred to the pattern recognition literature.[12]

V. GRAMMATICAL INFERENCE

In the discussions until now, we have assumed that the grammars generating the signals to be recognized are given. The automata used as recognizers or as translators require that the grammar be known. In most practical cases, however, the grammars are not given. Usually we are given a number of signal samples belonging to the same class. These are given by the "teacher" and are considered a part of the training set. Assuming all samples have been generated by the same grammar, it is now required to determine that grammar (or any other grammar that can generate the set). The problem is known as grammatical inference.[3] Since there is no unique one-to-one relationship between L(G) and G, grammatical inference may not provide unique results.

Grammatical inference is defined as follows. A finite set of sentences, S^+, belonging to L(G) is given. Possibly another set of sentences, S^-, belonging to $\overline{L(G)}$ (all sentences not belonging to L(G)) is also given. Using this information, it is required to infer the grammar, G. It is obvious that the inference procedure depends on the amount of information included in the set S^+. If S^+ includes all possible sentences, namely, $S^+ = L(G)$, it is called "complete". If S^+ is not complete, but each rewrite rule of G is used in the generation of at least one string in S^+, it is called "structurally complete".

Inference algorithms for finite-state, context-free, and stochastic context-free grammars have been developed. The subject is of increasing importance; its applications are found in the study of natural languages and in pattern and signal analysis.

VI. EXAMPLES

A. Syntactic Analysis of Carotid Blood Pressure

The detailed monitoring of arterial blood pressure is important for the care of the critically ill. Arterial blood pressure can be monitored by means of a pressure transducer inserted through a catheter into an artery. The waveforms monitored closely correlate with the heart dynamics. A typical blood pressure waveform is shown in Figure 12. The wave can be divided into two parts corresponding to the pressures of the systolic and diastolic phases of the heart. The main features of the pressure wave[28] are a high and rapid pressure rise at the beginning of systole reaching a peak (sometimes with "ringing") followed by a drop in pressure. The systolic phase ends with the closing of the aortic valve. The pressure then rises due to the compliance of the aorta. The minima on the pressure wave present at the end of systole is known as the dicrotic notch. (For more details on the blood pressure wave, see Appendix A.)

The carotid artery is a main artery supplying the brain. Its analysis is of special importance. Stockman et al.[23] have suggested syntactic method for the analysis. Their example has been used also by others.[7] Stockman and Kanal[5] have used a training set of 20 carotid pulse waves to check their parsing algorithm. Out of 158 waves analyzed, 125 were correctly recognized.

The primitives chosen to describe the signal were

LLP	A long line with large position slope
MLP	A medium-length line with large positive slope
MLN	A medium-length line with large negative slope
MP	A medium-length line with positive slope
MN	A medium-length line with negative slope

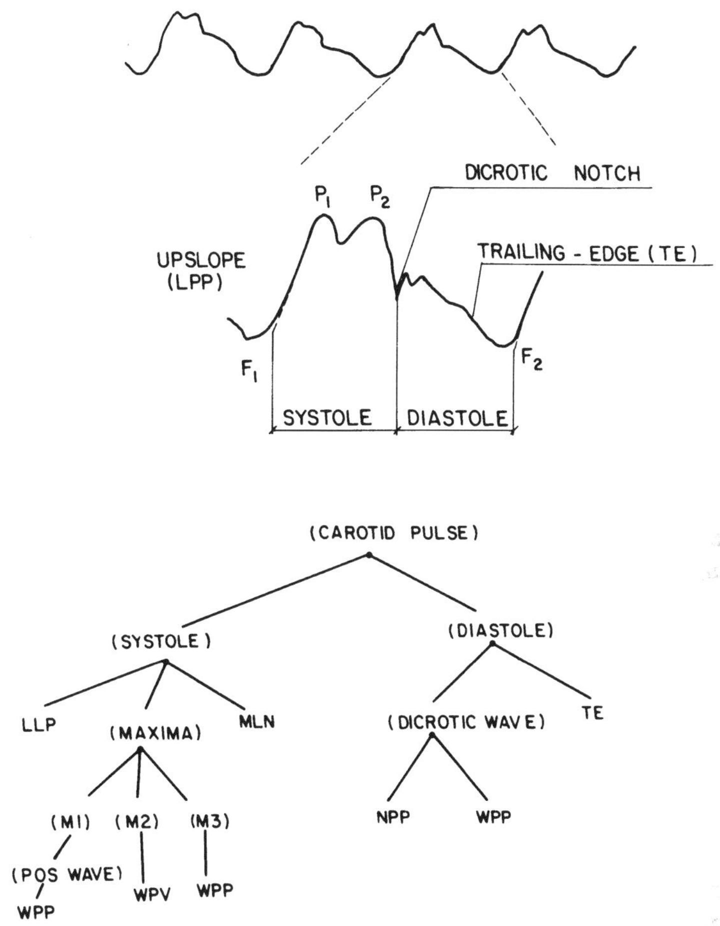

FIGURE 12. Carotid blood pressure wave with relational tree.

TE	Trailing edge — a long line with medium negative slope
HR	A short almost horizontal line
WPP	Wide parabola, peak
NPP	Narrow parabola, peak
WPV	Wide parabola, valley
NPV	Narrow parabola, valley
RPM	Right half of parabolic maxima
LPM	Left half of parabolic maxima

A typical *systole* part may contain the following primitives: LLP, WPP, WPV, WPP, MLN; and a typical *diastole* may contain NPP, WPP, TE.

A context-free grammar, G_p, has been chosen[5] to describe the signal with

$$G_p = (V_N, V_T, R, \text{(Carotid pulse)}) \tag{4.16}$$

where:

V_N = {(Carotid pulse), (Systole), (Diastole), (Maxima),·(M1), (M2), (M3), (Dicrotic wave), (Pos wave), (Neg wave)};

V_T = {LLP,MLP,MLN,MP,MN,TE,HR,WPP,NPP,WPV,NPV,RPM,LPM}

R:	(Carotid pulse)	→	(Systole)(Diastole)
	(Systole)	→	LLP(Maxima)MLN
	(Maxima)	→	(M1)(M2)(M3)
	(Maxima)	→	MP(M3)
	(Maxima)	→	(M1)MN
	(Diastole)	→	TE
	(Diastole)	→	(Dicrotic wave)TE
	(Dicrotic wave)	→	WPP
	(Dicrotic wave)	→	HR
	(Dicrotic wave)	→	NPP
	(Dicrotic wave)	→	NPP WPP
	(M1) → LPM;		(M1) → (Pos wave)
	(M2) → WPV;		(M2) → (Neg wave)
	(M3) → RPM;		(M3) → WPP
	(Pos wave)	→	WPP
	(Pos wave)	→	WPP MLN
	(Neg wave)	→	NPV
	(Neg wave)	→	NPV MLP

The following are a few waves belonging to $L(G_p)$: {LLP,MP,RPM,MLN,WPP,TE}, {LLP,LPM,WPV,RPM,MLN,HR,TE}, {LLP,WPP,MLN,MN,MLN,TE}.

B. Syntactic Analysis of ECG

Several syntactic algorithms have been suggested for the analysis of the ECG signal, especially for the problem of QRS complex detection.[16,19-22] A simple syntactic QRS detection algorithm, implemented on a small portable device, was suggested by Furno and Tompkins.[21] A simple finite state automaton, A_E, has been designed given by

$$A_E = (\Sigma_E, Q_E, \delta, q_0, \{q_Q, q_N\}) \tag{4.17}$$

where:

$$\Sigma_E = \{\text{normup, normdown, zero, other}\}$$
$$Q_E = \{q_o, q_1, q_2, q_Q, q_N\}$$

The two terminal states, q_Q and q_N, correspond to a QRS wave and noise. The state transition rules of A_E are

$$\delta(q_0,\text{normup}) \rightarrow \{q_1\}$$
$$\delta(q_0,\text{zero}) \rightarrow \{q_0\}$$
$$\delta(q_0,\text{other}) \rightarrow \{q_N\}$$
$$\delta(q_1,\text{normdown}) \rightarrow \{q_2\}$$
$$\delta(q_1,\text{other}) \rightarrow \{q_N\}$$
$$\delta(q_2,\text{normup}) \rightarrow \{q_Q\}$$
$$\delta(q_2,\text{other}) \rightarrow \{q_N\}$$

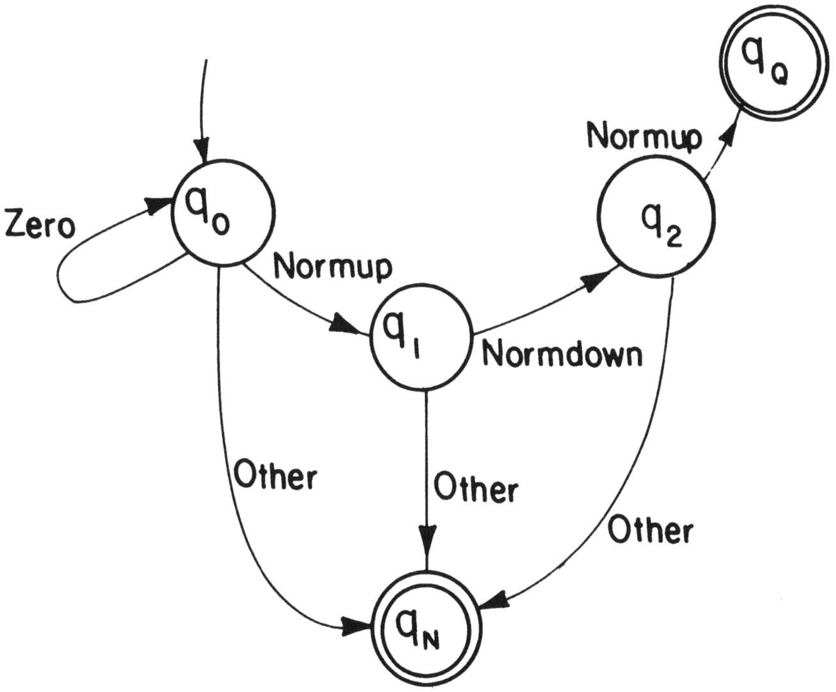

FIGURE 13. State transition diagram of finite state automation, A_E, for ECG analysis.

The state transition diagram of the automaton A_E is depicted in Figure 13. The primitives (normup, normdown, zero, and other) are calculated as follows. The ECG signal, x(t), is sampled with sampling interval, T. The derivative of x(t) is approximated by the first difference, s(k):

$$s(k) = \frac{x(kT) - x(kT - T)}{T} \tag{4.18}$$

The samples {s(k)} are grouped together into sequences. Each sequence consists of consecutive samples with the same sign. Consider, for example, the case where $s(n - 1) < 0$ and $s(n) > 0$.

A new sequence of positive first differences is generated:

$$\{s(n), s(n + 1), \ldots, s(m)\} \tag{4.19}$$

where $s(m + 1)$ is the first sample to become negative. Two numbers are associated with the sequence (Equation 4.19), the sequence length, S_L, and the sequence sum, S_M:

$$S_L = m - n + 1$$

$$S_M = \sum_{k=n}^{m} S(k) \tag{4.20}$$

Using predetermined thresholds on S_L and S_M, the primitives are extracted. The algorithm has been reported to operate at about ten times real time.

Table 1
PRIMITIVE EXTRACTION FOR QRS DETECTION

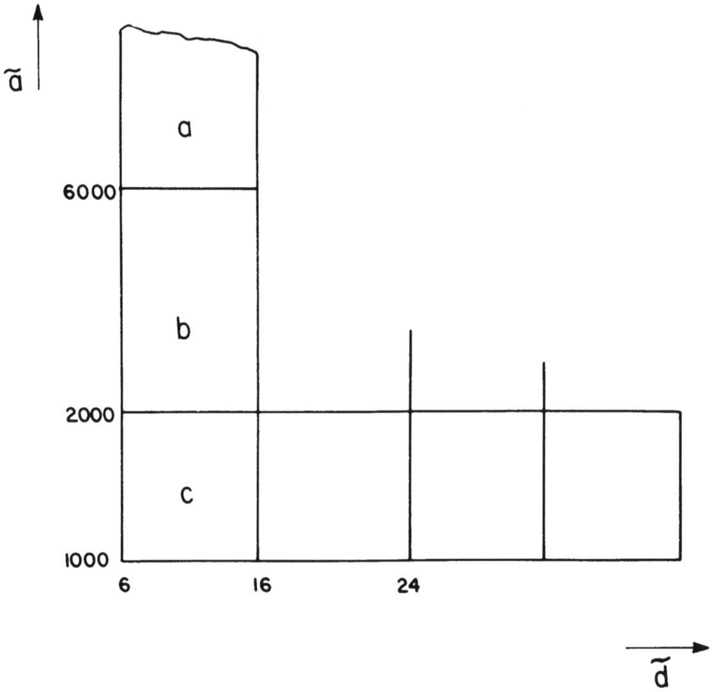

A more elaborate syntactic QRS detection algorithm has been suggested by Belforte et al.[19] Here a three-lead ECG was used. The first differences of the three signals were computed (Equation 4.18), yielding $s_i(k)$, $i = 1,2,3$. The energy of the first differences, $s_i^2(k)$, $i = 1,2,3$, were used to extract the primitives. A threshold was determined for the energy and the pulses above this threshold were considered. The peak of a pulse was denoted, $\tilde{a}$, and its duration (time above threshold) was denoted, $\tilde{d}$. The quantaties $\tilde{a}$ and $\tilde{d}$ were roughly quantized by means of Table 1, yielding the primitives a,b,c. Peaks were considered as belonging to different events every time the interval between them was longer than 80 msec. Strings were thus separated by the end of string symbol, w.

A sample of one lead of the ECG, derivative, and energy are shown in Figure 14. Pulses above threshold may belong to a QRS complex or may be the result of noise. A string, from lead i, that may be the result of a QRS complex is called a QRS hypothesis and is denoted Q_i. A grammar has been inferred from training samples that always appeared with QRS complexes. This grammar was denoted G_Q. Another grammar, G_Z, has been introduced representing strings that in the training set sometimes were from QRS complexes and sometimes were not. The two grammars are given by

$$G_Q = \{V_{NQ}, V_{TQ}, R_Q, QRS\} \tag{4.21}$$

where

$$V_{NQ} = \{U_1, U_2, U_3, U_4, QRS\}$$

$$V_{TQ} = \{a, b, c\}$$

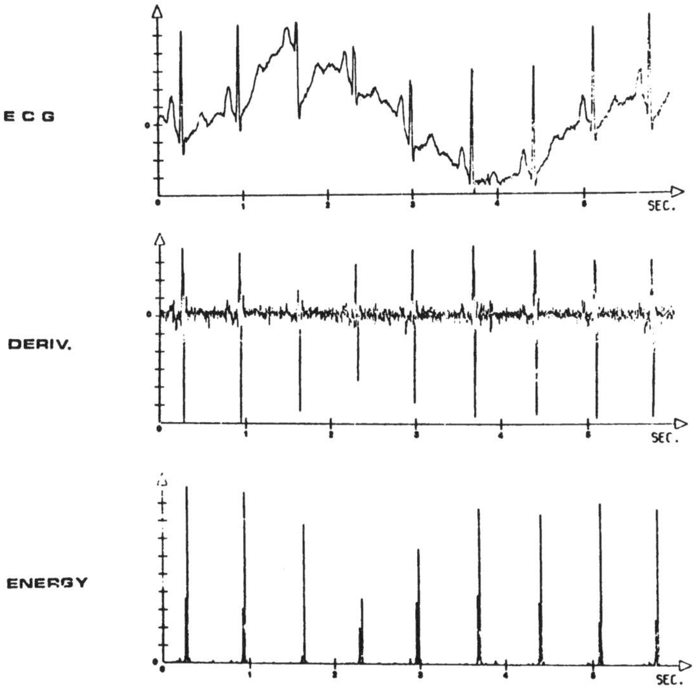

FIGURE 14. Syntactic QRS detection — ECG, derivative, and energy. (From Belforte, G., De-Mori, R., and Ferraris, F., *IEEE Trans. Biomed. Eng.*, BME-26, 125, 1979 (© 1979, IEEE). With permission.)

$$
\begin{array}{llll}
R_Q: & QRS \rightarrow bU_1; & QRS \rightarrow cU_2; & QRS \rightarrow aU_4 \\
& U_1 \rightarrow cU_2; & U_1 \rightarrow bU_3; & U_1 \rightarrow aU_4 \\
& U_2 \rightarrow bU_1; & U_2 \rightarrow cU_2; & U_2 \rightarrow aU_4 \\
& U_3 \rightarrow cU_2; & U_3 \rightarrow aU_4; & U_3 \rightarrow bU_4 \\
& U_4 \rightarrow aU_4; & U_4 \rightarrow bU_4; & U_4 \rightarrow cU_4 \\
& U_4 \rightarrow a; & U_4 \rightarrow b; & U_4 \rightarrow c
\end{array}
$$

and

$$G_z = \{V_{Nz}V_{Tz},R_z,Z\} \qquad (4.22)$$

where

$$
\begin{array}{lll}
V_{Nz} = \{Y_1,Y_2,Z\} & & \\
V_{Tz} = \{b,c\} & & \\
R_z: & Z \rightarrow cY_1 \ ; & Z \rightarrow bY_2 \\
& Y_1 \rightarrow cY_1 \ ; & Y_1 \rightarrow bY_2 \\
& Y_2 \rightarrow cY_2 \ ; & Y_2 \rightarrow b
\end{array}
$$

for example, the strings {bcbcaa}, {b^n}, and {bc^naa} are generated by G_Q and [cbb] and [bc^nb] are generated by G_z.

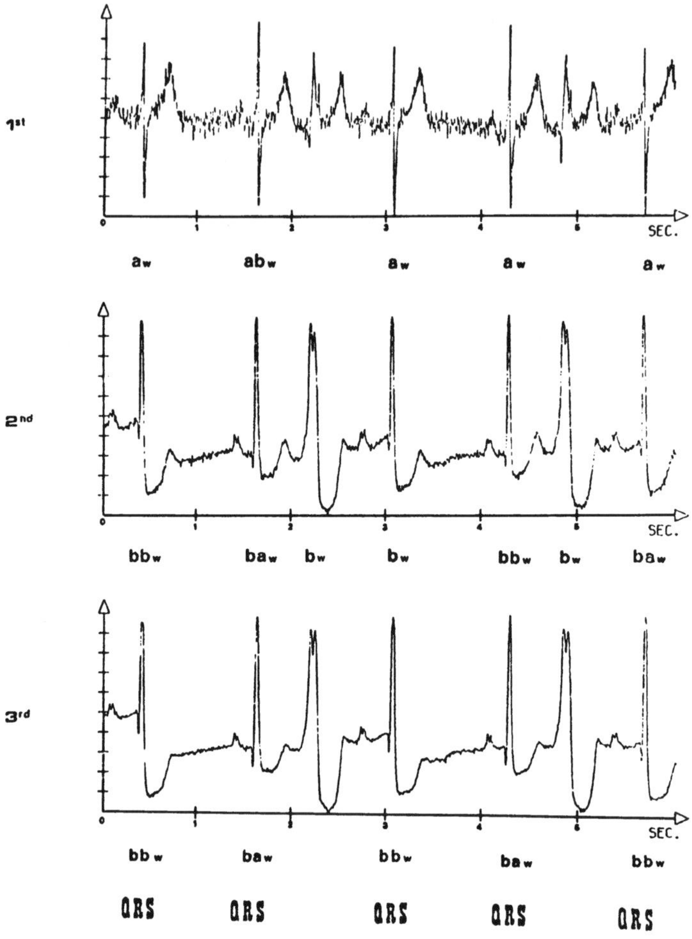

FIGURE 15. Syntactic QRS detection — three leads. (From Belforte, G., De-Mori, R., and Ferraris, F., *IEEE Trans. Biomed. Eng.*, BME-26, 125, 1979 (© 1979, IEEE). With permission.)

The rule suggested by Belforte et al.[19] for recognizing a QRS event is as follows. Let Q_i, $i = 1,2,3$, be a QRS hypothesis emitted under the control of grammer G_Q in the time interval $\{t_{i,1}, t_{i,2}\}$, where i denotes the lead number. Let also Z_j, $j = 1,2,3$, be the hypothesis emitted under the grammar G_z in the time $\{t_{j,1},t_{j,2}\}$. For a given lead and time interval, only one hypothesis can be emitted since the grammars G_Q and G_z generate disjoint languages.

The hypothesis Q_i and Z_j, $i,j = 1,2,3$, whose time intervals partially overlap are used to determine the presence or absence of a QRS complex. The decision rule suggested[19]

$$h = Q_i \Lambda (Q_j V z_j) \quad ; \quad i,j = 1,2,3 \qquad i \neq j \tag{4.23}$$

where Λ and V are the logical "and" and "inclusive or" operators. A QRS is declared if $h = 1$. The algorithm was checked, in real time, with data base of 620 QRSs from 16 healthy and ill patients with no errors and less than 0.5% false alarm errors. Examples of the three-lead ECG and detection results are shown in Figure 15.

C. Syntactic Analysis of EEG

In the analysis of EEG, spatiotemporal information is of considerable importance. Syntactic methods may have a good potential for EEG analysis since they utilize this information. Syntactic analysis of EEG spectra has been suggested.[11-14]

The EEG was divided into nonoverlapping segments of 1-sec duration. The spectrum of each epoch was estimated (by AR modeling). Discriminant analysis of the training set generated seven discriminant functions:

$$\{AL,A,SL,S,L,NL,N\} \tag{4.24}$$

with AL = artifactual low, A = artifactual, SL = slow low, S = slow, L = low, NL = normal low, and N = normal. These were defined as the seven primitives. Recognizable entities in the EEG such as normal, abnormal, drowsy, lowamp, waxing and waning, or slow were used as nonterminal states. Rewrite rules were parsed from the training set.

EEG records (from healthy population), analyzed as normal by expert evaluation, were all (55 records of 9 sec each) classified normal by the syntactic algorithm. In EEG records drawn from a dialysis population, results were somewhat less successful. From the records classified abnormal by the expert, 11% were classified normal by the syntactic algorithm. From the records classified normal by the expert, about 2% were classified abnormal by the algorithm.

REFERENCES

1. **Fu, K. S.,** *Syntactic Methods in Pattern Recognition*, Academic Press, New York, 1974.
2. **Gonzalez, R. C. and Thomason, M. G.,** *Syntactic Pattern Recognition, An Introduction*, Addison-Wesley, London, 1978.
3. **Fu, K. S. and Booth, T. L.,** Gramatical inference: introduction and survey. I and II, *IEEE Trans. Syst. Man Cybern.*, 5, 95, 409, 1975.
4. **Pavlidis, T.,** Linguistic analysis of waveforms, in *Software Engineering*, Vol. 2, Tou, J. T., Ed., Academic Press, 1971, 203.
5. **Stockman, G. C. and Kanal, L. N.,** Problem reduction representation for the linguistic analysis of waveforms, *IEEE Trans. Pattern Anal. Mach. Intelligence*, 5, 287, 1983.
6. **Mottl, V. V. and Muchnik, I. B.,** Linquîstic analysis of experimental curves, *Proc. IEEE*, 67, 714, 1979.
7. **Fu, K. S.,** Syntactic pattern recognition and its applications to signal processing, in *Digital Waveform Processing and Recognition*, Chen, C. H., Ed., CRC Press, Boca Raton, Fla., 1982, chap. 5.
8. **Sankar, P. V. and Rosenfeld, A.,** Hierarchical representation of waveforms, *IEEE Trans. Pattern Anal. Mach. Intelligence*, 1, 73, 1979.
9. **Ehrich, R. W. and Foith, J. P.,** Representation of random waveforms by relational trees, *IEEE Trans. Comput.*, 25, 725, 1976.
10. **Lin, H. H. and Fu, K. S.,** An application of syntactic pattern recognition to siesmic discrimination, *IEEE Trans. Geosci. Remote. Sens.*, 21, 125, 1983.
11. **Bourne, J. R., Jagannathan, V., Hammel, B., Jansen, B. H., Ward, J. W., Hughes, J. R., and Erwin, C. W.,** Evaluation of a syntactic pattern recognition approach to quantitative EEG analysis, *Electroencephalogr. Clin. Neurophysiol.*, 52, 57, 1981.
12. **Bourne, J. R., Gagannathan, V., Giese, B., and Ward, J. W.,** A software system for syntactic analysis of the EEG, *Comput. Prog. Biomed.*, 11, 190, 1980.
13. **Giese, D. A., Bourne, J. R., and Ward, J. W.,** Syntactic analysis of the electroencephalogram, *IEEE Trans. Syst. Man Cybern.*, 9, 429, 1979.
14. **Jansen, B. H., Bourne, J. R., and Ward, J. W.,** Identification and labeling of EEG graphic elements using autoregressive spectral estimates, *Comput. Biol. Med.*, 12, 97, 1982.
15. **Albus, J. E.,** ECG interpretation using stochastic finite state model, in *Syntactic Pattern Recognition Applications*, Fu, K. S., Ed., Springer-Verlag, Berlin, 1976.
16. **Horowitz, S. L.,** A syntactic algorithm for peak detection in waveforms with applications to cardiography, *Commun. ACM*, 18, 281, 1975.
17. **Degani, R. and Pacini, G.,** Fuzzy classification of electrocardiograms, in *Optimization of Computer ECG Processing*, Wolf, H. K. and MacFarlane, P. W., Eds., North-Holland, Amsterdam, 1980, 217.

18. **Smets, P.,** New quantified approach for diagnostic classification, in *Optimization of Computer ECG Processing,* Wolf, H. K. and MacFarlane, P. W., Eds., North-Holland, Amsterdam, 1980, 229.
19. **Belforte, G., De-Mori, R., and Ferraris, F.,** A contribution to the automatic processing of electrocardiograms using syntactic methods, *IEEE Trans. Biomed. Eng.,* 26, 125, 1979.
20. **Papakonstantinou, G. and Gritzali, F.,** Syntactic filtering of ECG waveforms, *Comput. Biomed. Res.,* 14, 158, 1981.
21. **Furno, G. S. and Tompkins, W. J.,** QRS detection using automata theory in a battery powered microprocessor system, *Proc. IEEE Frontiers Eng. Health Care,* 155, 1982.
22. **Birman, K. P.,** Using SEEK for Multi Channel Pattern Recognition, Tech. Rep. 82-529, Department of Computer Science, Cornell University, Ithaca, N.Y., 1982.
23. **Stockman, G., Kanal, L., and Kyle, M. C.,** Structural pattern recognition of carotid pulse waves using general waveform parsing system, *Commun. ACM,* 19, 688, 1976.
24. **De-Mori, R.,** *Computer Models of Speech Using Fuzzy Algorithms,* Plenum Press, New York, 1983.
25. **Pavlidis, T. and Horwitz, S. L.,** Segmentation of plane curves, *IEEE Trans. Comput.,* 23, 860, 1974.
26. **Tomek, I.,** Two algorithms for piecewise linear continuous approximation of functions of one variable, *IEEE Trans. Comput.,* 23, 445, 1974.
27. **Pavlidis, T.,** *Structural Pattern Recognition,* Springer-Verlag, Berlin, 1980.
28. **Cox, J. R., Nolle, F. M., and Arthur, R. M.,** Digital analysis of the electroencephalogram, the blood pressure wave and the electrocardiogram, *Proc. IEEE,* 60, 1137, 1972.
29. **Birman, K. P.,** Rule based learning for more accurate ECG analysis, *IEEE Trans. Pattern Anal. Mach. Intelligence,* 14, 369, 1982.

Appendix A

CHARACTERISTICS OF SOME DYNAMIC BIOMEDICAL SIGNALS

I. INTRODUCTION

The typical levels and frequency ranges of various biomedical signals are briefly discussed in this appendix. Only rough ranges are given because of the large variances that exist in these types of signals, and the strong dependence on the acquisition method. Records of typical signals are shown for most of the signals discussed here. A brief discussion on the main processing methods and problems is presented. Because of the large amounts of information available concerning the effects of various abnormalities on the signals, especially on the more important ones such as the ECG or EEG, it was impossible to present a detailed discussion. Selected references are given that refer the reader to a more detailed discussion for each signal. The signals have been divided into groups according to inherent characteristics. In some cases, however, the division is not perfectly clear.

II. BIOELECTRIC SIGNALS

A. Action Potential

This is the potential generated by the excitable membrane of a nerve or muscle cell (Chapter 2, Volume I). The *action potential* generated by a single cell can be measured by means of a microelectrode inserted into the cell and a reference electrode located in the extracellular fluid. The microelectrode has a very high input impedance. An amplifier with a very low noise figure and input capacitance must be used.[1]

In most applications the shape of the action potential is of no interest. It is the interspike intervals that are of interest (Figure 1, Chapter 2). The time of occurrence of the spike is detected and point process methods are used (Chapter 2).

When the action potentials from more than one unit are monitored by the electrode, multispike[2] train analysis techniques are required. The action potentials from the various neurons can be identified by template matching methods (Chapter 1) and marked point processes analysis can be applied. Typical level range of the action potential is 100 mV. The bandwidth required is about 2 kHz.

B. Electroneurogram (ENG)

The field generated by a nerve can be measured without penetrating the membrane of a single cell. A needle electrode inserted to the nerve bundle or even surface electrodes located on the skin can measure the signal. The monitored voltage[3] will not, in general, be a single action potential, but the contribution of several action potentials transmitted through the volume conductor.

Figure 1 depicts the ENG recorded by surface electrodes from the median nerve. The range of levels is about 5 μV to 10 mV with a bandwidth of about 1 kHz. The ENG is used clinically to calculate nerve conduction velocity. This information is required for the detection of nerve fiber damage or regeneration. Because of the low amplitudes involved in the ENG monitoring, synchronized averaging methods (Chapter 5, Volume I) are often used to increase the signal to noise ratios.

C. Electroretinogram (ERG)

The ERG is the potential generated by the retina.[4-7] Evoked ERG is most often used, which is the potential generated by a short flash of light. The ERG is used clinically[6] and

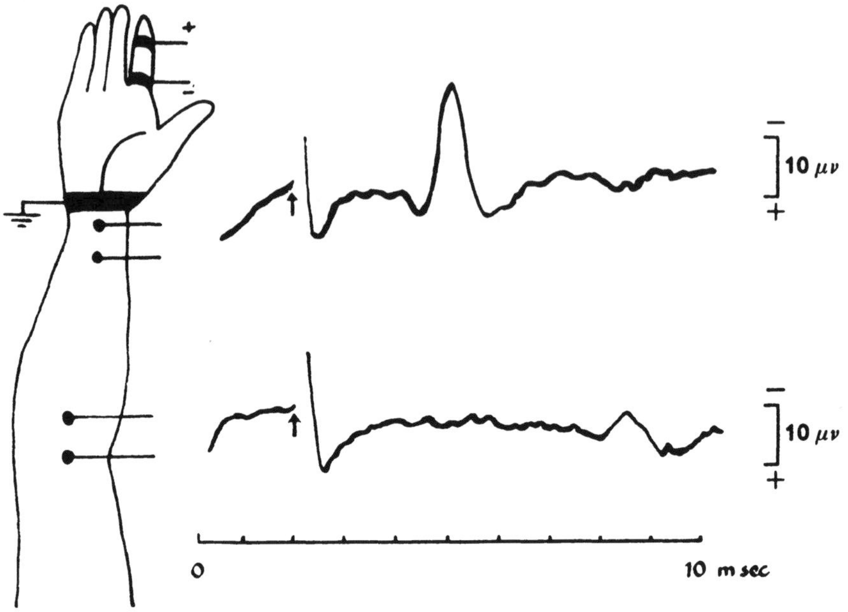

FIGURE 1. Sensory nerve action potentials evoked from the median nerve at the elbow and wrist after stimulation of the index finger. (From Lenman, J. A. R. and Ritchie, A. E., *Clinical Electromyography*, Pitman Medical and Scientific, London, 1970. With permission.)

in ophthalmological research. For research purposes, it is acquired by an implanted microelectrode in the retina and an indifference electrode elsewhere on the surface. For clinical use, a cornea electrode (usually a specially made contact lens) is employed. The indifference electrode is placed on the earlobe, the temple, or the forehead.

The ERG is typically composed of four components, the "a" (negative), "b" (positive), "c" (positive), and "d" (negative) waves. The various waves are assumed to be generated by different regions of the cornea. The a and c waves are probably contributed by the deepest part of the cornea. The b wave probably originates in the bipolar cells region. The d wave is associated with the termination of the stimulus.

The voltage levels of the ERG are in the range of 0.5 μV to 1 mV in clinical applications. Much higher voltages are acquired in research experiments when the electrode is implanted in the retina. The duration of the a, b, and c waves is about 0.25 to 1.5 sec, with the d wave appearing at the termination of the stimulus. The bandwidth required for the processing of the ERG is about 0.2 to 200 Hz. Processing techniques applied to the ERG are mainly synchronized averaging. Nonlinear methods have also been applied.[7]

D. Electro-Oculogram (EOG)

The EOG is the recording of the steady *corneal-retinal potential.*[8-11] This potential has been used to measure eye position, either for research purposes (sleep research) or for clinical use. The signal is measured by pairs of surface electrodes placed to the left and right of the eyes and above and below the eyes. The amplitude levels are in the range of 10 μV to 5 mV. The signal requires the frequency range of DC to 100 Hz.

E. Electroencephalogram (EEG)

The recording of the electrical activity of the brain is known as *electroencephalography* (EEG). It is widely used[11,12] for clinical and research purposes. Methods have been developed to investigate the functioning of the various parts of the brain by means of the EEG. Three types of recordings are used. *Depth recording* is done by the insertion of needle electrodes into the neural tissue of the brain. Electrodes can be placed on the exposed surface of the brain, a method known as *electrocorticogram.* The most generally used method is the noninvasive recording from the scalp by means of surface electrodes.

The investigation of the electrical activity of the brain is generally divided into two modes. The first is the recordings of spontaneous activity of the brain which is the result of the electrical field generated by the brain with no specific task assigned to it. The second is the *evoked potentials* (EP). These are the potentials generated by the brain as a result of a specific stimulus (such as a flash of light, an audio click, etc.). EPs are described in the next section.

The surface recording of the EEG depends on the locations of the electrodes. In routine clinical multiple EEG recordings, the electrodes are placed in agreed upon locations in the frontal (F), central (C), temporal (T), parietal (P), and occipital (O) regions, with two common electrodes placed on the earlobes. Between 6 to 32 channels are employed, with 8 or 16 being the number most often used. Potential differences between the various electrodes are recorded. There are three modes of recordings: the unipolar, averaging reference, and bipolar recordings (e.g., see Strong).[17]

The bandwidth range of the scalp EEG is DC to 100 Hz, with the major power distributed in the range of 0.5 to 60 Hz. Amplitudes of the scalp EEG range from 2 to 100 μV. The EEG power spectral density varies greatly with physical and behavioral states. EEG frequency analysis has been a major processing tool in neurological diagnosis for many years. It has been used for the diagnosis of epilepsy, head injuries, psychiatric malfunctions, sleep disorders, and others. The major portion of the EEG spectrum has been subdivided into fine bands.

The delta range — The part of the spectrum that occupies the frequency range of 0.5 to 4 Hz is the delta range. Delta waves appear in young children, deep sleep, and in some brain diseases. In the alert adult, delta activity is considered abnormal.

The theta range — The theta range is the part of the spectrum that occupies the frequency range of 4 to 8 Hz. Transient components of theta activities have been found in normal adult subjects in the alert state. The theta activity occurs mainly in the temporal and central areas and is more common in children.

The alpha range — The alpha range is the part of the spectrum that occupies the range of 8 to 13 Hz. These types of rhythms are common in normal subjects, best seen when the subject is awake, with closed eyes, under conditions of relaxation. The source of the alpha waves is believed to be in the occipital lobes. An example of the alpha activity can be seen in Figure 2.

The beta range — The beta range is the part of the spectrum that occupies the range 13 to 22 Hz. The beta rhythms are recorded in the normal adult subject mainly from the precentral regions, but many appear in other regions as well. The beta range has been subdivided into two: Beta I is the higher frequency range and beta II is the lower frequency range. Beta II is present during intense activation of the CNS, while beta I is diminished by such activation. Sedatives and various barbiturates cause an increase[11] of beta activity often up to amplitudes of 100 μV.

Time domain analysis is also used for EEG processing to detect short wavelets. This has been applied mainly in sleep analysis. Sleep is a dynamic process which consists of various stages. At the beginning of the process the subject is in a state of drowsiness where widespread alpha activity appears. Light sleep, stage 1, is characterized by low voltages of mixed frequences. Sharp waves may appear in the EEG. These are the result of a response to stimuli and are known as *V-waves* (Figure 3b). The spectrum at stage 1 of sleep is dominated by theta waves. In state 2, the slow activity is increased and *sleep spindles* appear. These are bursts of about 3 to 5 cycles of alpha-like activity with amplitude of about 50 to 100 μV. In stages 3 (moderate sleep) and 4 (deep sleep), there is an increase in irregular

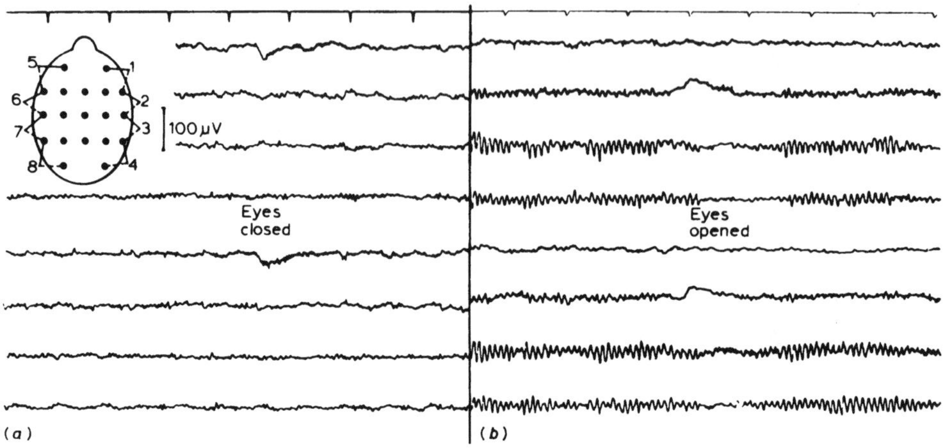

FIGURE 2. EEG recordings. (a) Subject with complete absence of alpha waves; (b) subject with alpha waves, diminished for only about 1 sec following eye opening. (From Kiloh, L. G., McComas, A. J., Osselton, J. W., and Upton, A. R. M., *Clinical Electroencephalography*, 4th ed., Butterworths, London, 1981. With permission.)

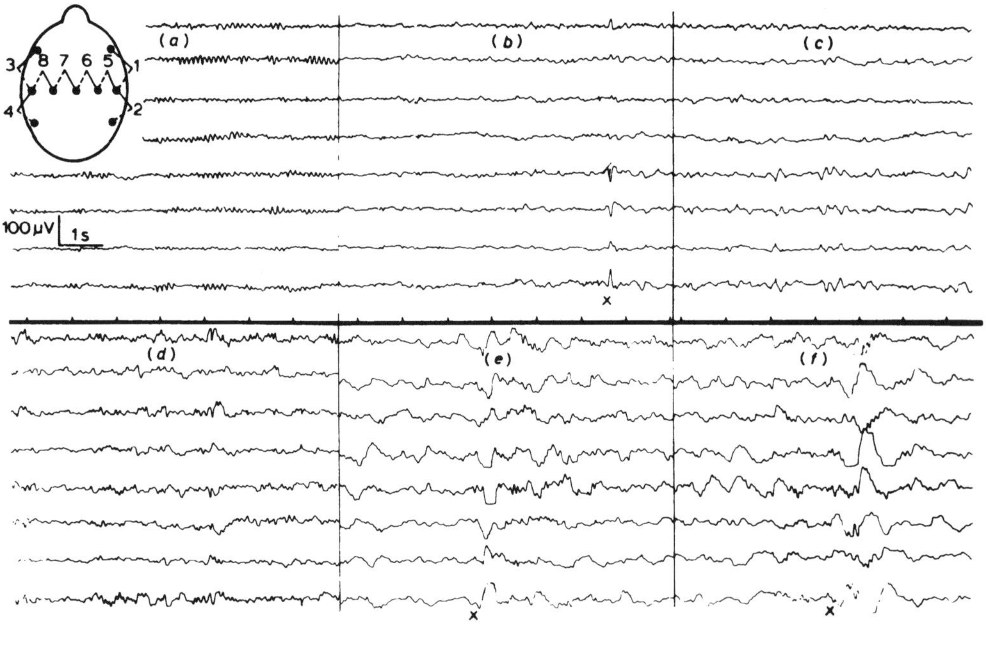

FIGURE 3. EEG recordings, stages of drowsiness and sleep. (a) Early drowsiness, widespread alpha rhythm; (b) light sleep (stage 1), note vertex sharp waves in response to sound stimulus at X; (c) light sleep, theta dominant stage; (d) stage 2, emerging of sleep spindles; (e) and (f) stages 3 and 4. Increasing irregular delta activity, K-complex responses to sound stimuli at X. (From Kiloh, L. G., McComas, A. J., Osselton, J. W., and Upton, A. R. M., *Clinical Electroencephalography*, 4th ed., Butterworths, London, 1981. With permission.)

delta activity and the appearance of *K-complexes*. These complexes, most readily evoked by an auditory stimulus, consist of a burst of one or two high-voltage (100 to 200 μV) slow waves, sometimes accompanied or followed by a short episode of 12- to 14-Hz activity[11] (Figure 3). Another sleep stage has been defined, the *rapid eye movements* (REM) stage. The EEG of the REM stage is similar to that of stage 1 and early stage 2, but in which REM appear. It has been also termed the *paradoxial* sleep state (Figure 4).

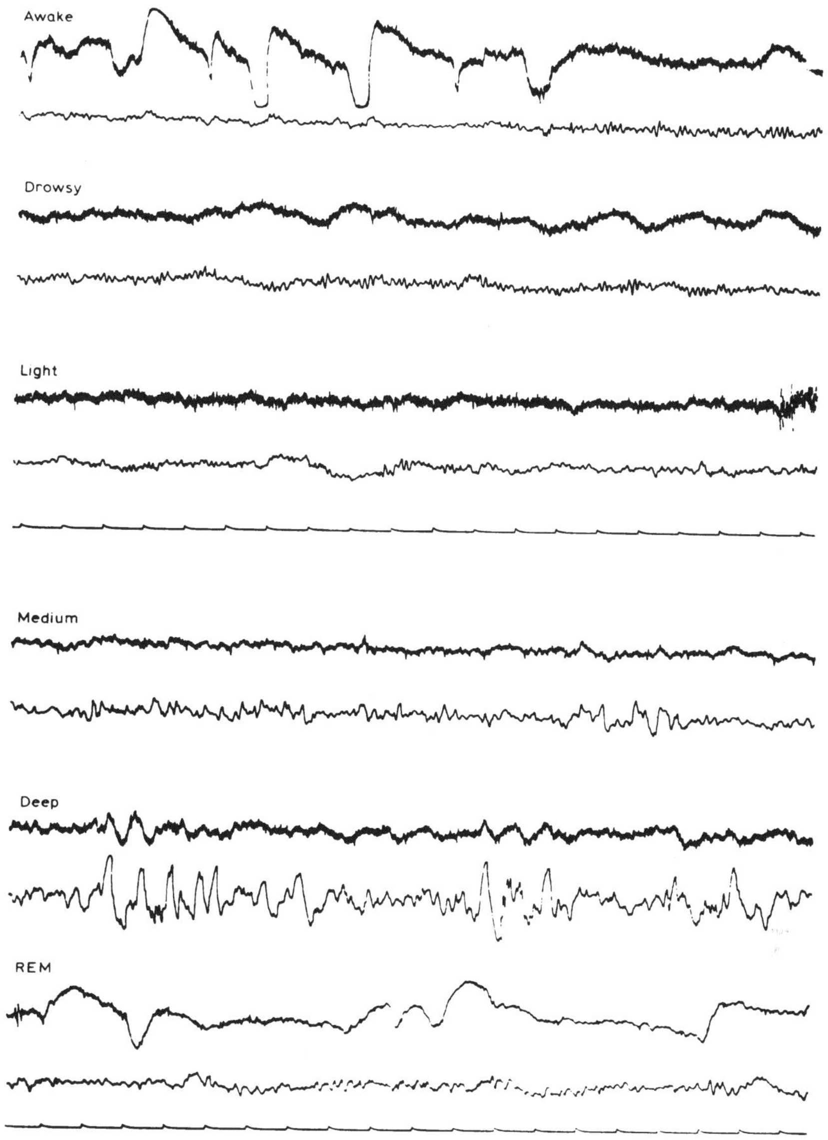

FIGURE 4. Stages of wakefulness and sleep. Upper channel of each pair, eye movements plus submental EMG; Lower channel, EEG. (From Kiloh, L. G., McComas, A. J., Osselton, J. W., and Upton, A. R. M., *Clinical Electroencephalography*, 4th ed., Butterworths, London, 1981. With permission.)

Several abnormalities are seen by the EEG. Epilepsy is a condition where uncontrolled neural discharges take place in some location in the CNS. Such a seizure unvoluntarily activates various muscles and other functions while inhibiting others. Several types of epilepsies are known, among them are the *grand* and *petit mal, myoclonic epilepsy,* and others (Figure 5).

F. Evoked Potentials (EP)

The electrical activity of the brain evoked by a sensory stimulus[11,18] is known as the *evoked potentials* (EP) or *evoked responses* (ER). It is usually measured over the sensory region of the brain corresponding to the stimulating modality. A sensory stimulus results in two kinds of potential changes in the EEG. The nonspecific response is a low-voltage

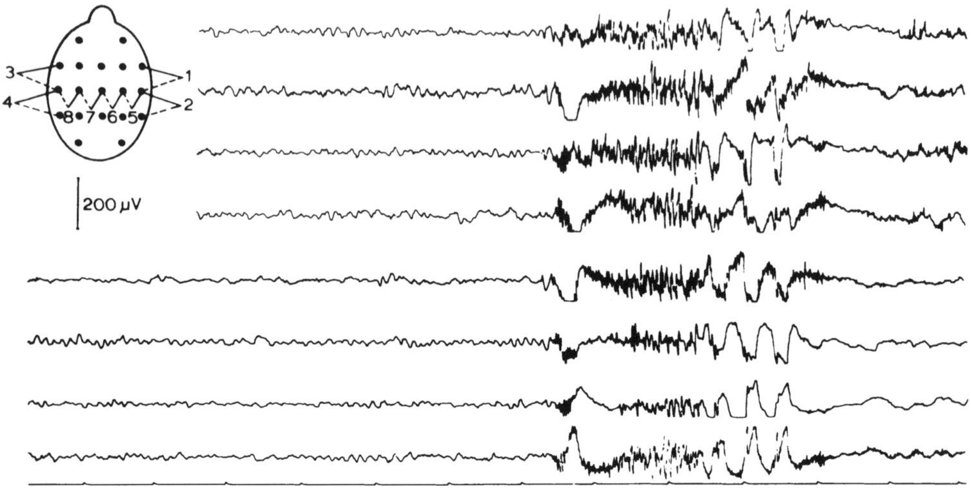

FIGURE 5. Generalized epilepsy. (From Kiloh, L. G., McComas, A. J., Osselton, J. W., and Upton, A. R. M., *Clinical Electroencephalography*, 4th ed., Butterworths, London, 1981. With permission.)

transient, having its maximum value in the region of the vertex. The response is similar in all types of stimuli. It becomes less marked when the same stimulus is repeated. The V-wave and K-complex discussed in the previous section are nonspecific EPs. The specific response is initiated with some latency after the stimulus has been applied. It has its maximum in the cortical area, appropriate to the modality of stimulation.

The EP is very low in amplitude, which is in the range of 0.1 to 10 μV. The ongoing EEG in which the EP is burried may be an order of magnitude larger. Synchronized averaging techniques are usually used to detect the *average evoked potential* (AEP) (an abbreviation used also for *auditory evoked potentials)*. When the single EP is required,[19] other methods of signal to noise enhancement must be used (Chapter 1). There are essentially three major types of evoked potentials in common use.

Visual evoked potential (VEP) — The VEP is recorded[19] from the scalp over the occipital lobe. The stimuli are light flashes or visual patterns. The VEP has an amplitude range of 1 to 20 μV with a bandwidth of 1 to 300 Hz. The duration of the VEP is of 200 msec. VEP has been used for the diagnosis of multiple sclerosis (the optical nerve is commonly affected by the disease), to check color blindness, to assess visual fields deficits, and to check visual acuity. Figure 6 shows a typical VER.

Somatosensory evoked potential (SEP, SSEP) — The SEP is recorded[23] with surface electrodes placed over the sensory cortex. The stimulus may be electrical or mechanical. The duration of cortical SEP is about 25 to 50 msec, with a bandwidth of 2 to 3000 Hz. Subcortical SEP is much longer and lasts about 200 sec. Figure 7 depicts cortical and subcortical SEPs. SEP is used to provide information concerning the dorsal column pathway between the periferal nerve fibers and the cortex.[11]

Auditory evoked potential (AEP) — AEPs are recorded by electrodes placed at the vertex.[21,22] The auditory stimulus can be a click, tone burst, white noise, and others. The AEP is divided[11] into the first potential (latency of about a millisecond), the early potential (eighth nerve and brainstem, 8 msec), the middle potential (8 to 50 msec), and the late potential (50 to 500 msec). The initial 10-msec response has been associated with brainstem activities. These brainstem auditory evoked potentials (BAEP) are very low in amplitude (about 0.5 μV). The AEP has a bandwidth of 100 to 3000 Hz. AEPs have been used to check hearing deficiencies, especially in children. Figure 7 depicts a typical cortical and subcortical AEP.

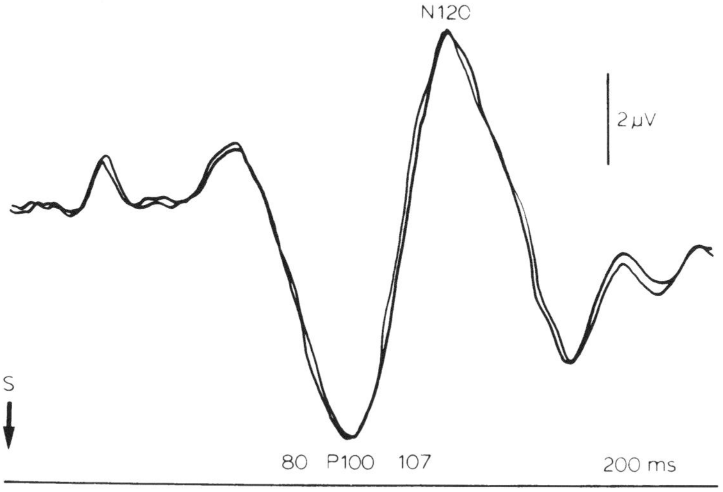

FIGURE 6. Averaged visual responses; bandwidth: 2 to 300 Hz; average of 64 responses. (From Kiloh, L. G., McComas, A. J., Osselton, J. W., and Upton, A. R. M., *Clinical Electroencephalography*, 4th ed., Butterworths, London, 1981. With permission.)

Other evoked potentials — Potentials evoked by pain[25] stimuli have been recorded. Such a stimulus can be an intense thermal pulse from an IR laser beam (Figure 7B). Olfactory evoked potentials have been reported as well as vestibulospinal potentials.

The processing of the EEG and EP requires many of the methods discussed in this book. The EEG is usually recorded in several channels for relatively long periods of time. Large amounts of data are thus collected. Automatic analysis and data compression techniques are needed. Time series analysis methods (Chapter 7, Volume I) have widely been applied to EEG analysis. Most often the EEG is modeled by an AR model,[26] and adaptive segmentation methods[27] are employed. The estimation of the EEG power spectral density (Chapter 8, Volume I) is an important part in both clinical and research oriented EEG analysis. Automatic classification methods (Chapter 3) have been applied to the EEG for automatic sleep staging, depth of anesthesia monitoring, and others. Wavelet detection methods (Chapter 1) have been used to detect K-complexes and spindles in the ongoing EEG. Principal components and singular value decomposition methods (Chapter 3) have been used[28] to analyze evoked potentials.

G. Electromyography (EMG)

EMG is the recording of the electrical potential generated by the muscle.[3,29] The activity of the muscle can be monitored by means of surface electrodes placed on the skin. The signal received yields information concerning the total electrical activity associated with the muscle contraction. More detailed information is often needed for clinical diagnosis. Concentric needle electrodes are then inserted through the skin into the muscle. The signal received is known as the *motor unit action potential* (MUAP). Higher resolution can be achieved by the use of microelectrodes by means of which single muscle fiber action potentials are recorded. The three types of EMG signals are briefly discussed here.[30]

Single fiber electromyography (SFEMG) — The action potentials recorded from a single muscle fiber have a duration of about 1 msec, with amplitudes of a few millivolts. The bandwidth used to process the SFEMG is 500 Hz to 10 kHz. Although the SFEMG contains low frequencies, it is advisable to cut off the low band so that contributions from more distant fibers (having most of their power in the low range due the volume conductor) can

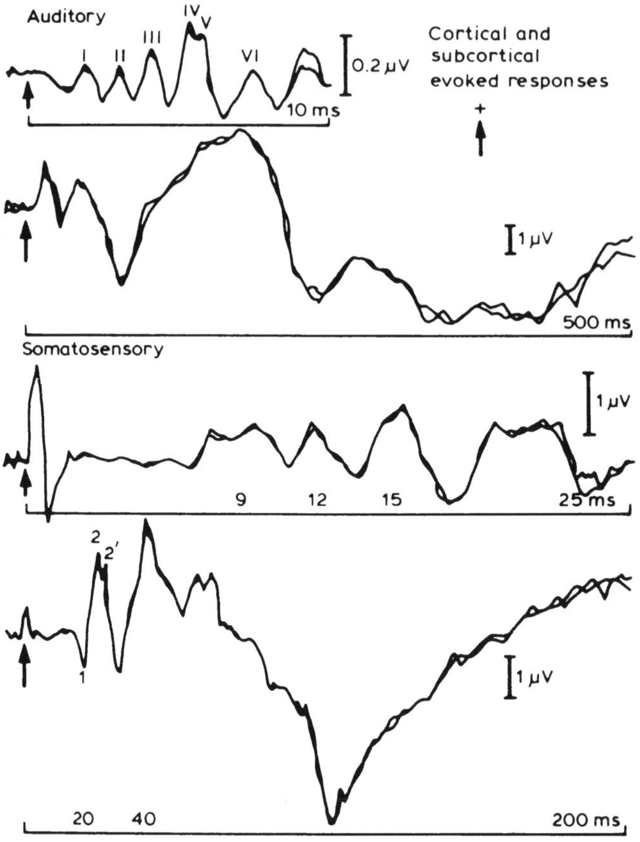

FIGURE 7A. Subcortical auditory (10 msec) evoked responses compared with equivalent cortical evoked responses (200 to 500 msec); subcortical: bandwidth 150 to 1500 Hz, 2000 averages; cortical: bandwidth 2 to 75 Hz, 64 averages. (From Kiloh, L. G., McComas, A. J., Osselton, J. W., and Upton, A. R. M., *Clinical Electroencephalography,* 4th ed., Butterworths, London, 1981. With permission.)

be minimized. SFEMG is used clinically to detect neuromuscular malfunctions such as myasthenia gravis.

Motor unit action potentials (MUAP) — The complex that consists of the nerve cell, the nerve fiber, the neuromuscular junctions, and the muscle fibers is called the *motor unit.* Action potentials recorded from this complex by means of concentric needle electrode are known as *motor unit action potentials* (MUAP, MUP). The MUAP is subdivided into two components. The spike is generated by about 2 to 12 fibers with a shape and amplitude that depend on the synchronicity among fibers within the MU. The initial and terminal parts are generated by more distant fibers of the same MU. The duration of the MUAP is about 2 to 10 msec, with amplitudes in the range of 100 µV to 2 mV. The bandwidth required for processing the MUAP is 5 Hz to 10 kHz. Figure 8 depicts a train of MUAP. The MUAP are clinically used to detect myopathies, neurogenic lesions, and other neuromuscular disorders. The processing[30] of the SFEMG and MUAP relies on signal to noise enhancement by means of synchronized averaging (Chapter 5, Volume I) and wavelet detection[31] (Chapter 2). Decomposition, classification, and dimension reduction methods[31-34] have also been applied to the analysis (Chapter 3).

Surface electromyography (SEMG) — The noninvasive acquisition of the EMG by means of surface electrodes is a convenient method. It does, however, yield only gross

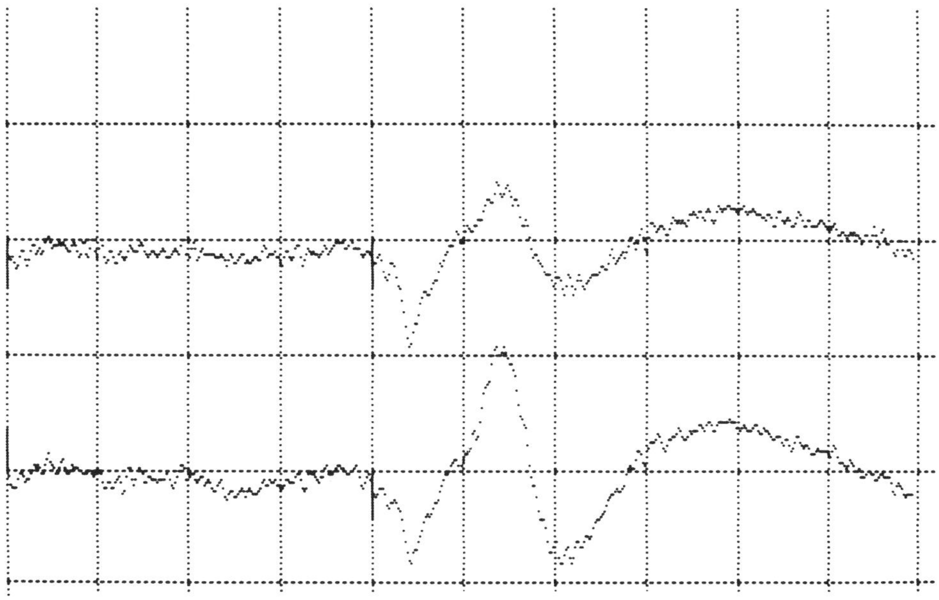

FIGURE 7B. Potential evoked from pain stimuli (intense laser beam pulse).

information on the muscle under investigation. The amplitudes of the SEMG depend on the muscle under investigation and the electrodes. The range of 50 μV to 5 mV is normal. The bandwidth required (for skeletal muscles) is 2 to 500 Hz (note that for smooth muscle the bandwidth required is 0.01 to 1 Hz). Figure 9 depicts a record of surface respiratory EMG. The power spectral density of the EMG[35-38] is often estimated for various clinical applications, such as pathological tremors[38] and fatigue analysis.[39,40] Surface EMG is also used for the control of prosthetics.[41-43] Time series analysis is widely applied to the SEMG both for power spectral density estimation[37] and for classification.[41] Linear discriminant functions have successfully been applied to the automatic classification of SEMG.[43]

H. Electrocardiography (ECG, EKG)

1. The Signal

The ECG is the recording of the electrical activity of the heart. The mechanical activity of the heart function is linked with the electrical activity. The ECG therefore is an important diagnostic tool for assessing heart function. Typical ECG signals are shown in Figure 1 of Chapter 1 and Figure 10, this chapter.

The heart electric cycle begins at the *sino atria* (SA) node on the right atrium; it is an astable bundle of nerves. The SA node paces the heart. SA node impulses cause the contraction of the atria, generating the p wave in the ECG. The impulses travel along conduction fibers within the atria to the *atrio ventricular* (AV) node which controls impulse transmission between atria and ventricles. The atrio ventricular conduction time is of the order of 120 to 220 msec. A special conduction system, consisting of the *bundles of His* and *Purkinje,* transfers the impulses into the lower and outer parts of the ventricles. The contraction of the ventricles comprises the pumping action of the heart and generates the QRS complex in the ECG. About 150 msec later the ventricles repolarize, causing the T wave in the ECG. The repolarization of the atria is rarely seen in the ECG. In the rare cases in which it does occur, it appears between the P and Q waves and is called the TA wave. An additional wave, the U wave, is sometimes recorded after the T wave. Its cause is believed to be the repolarization of ventricular papillary muscles.

The cardiac rhythm, or heart beat rate, is a random process. It is usually measured by

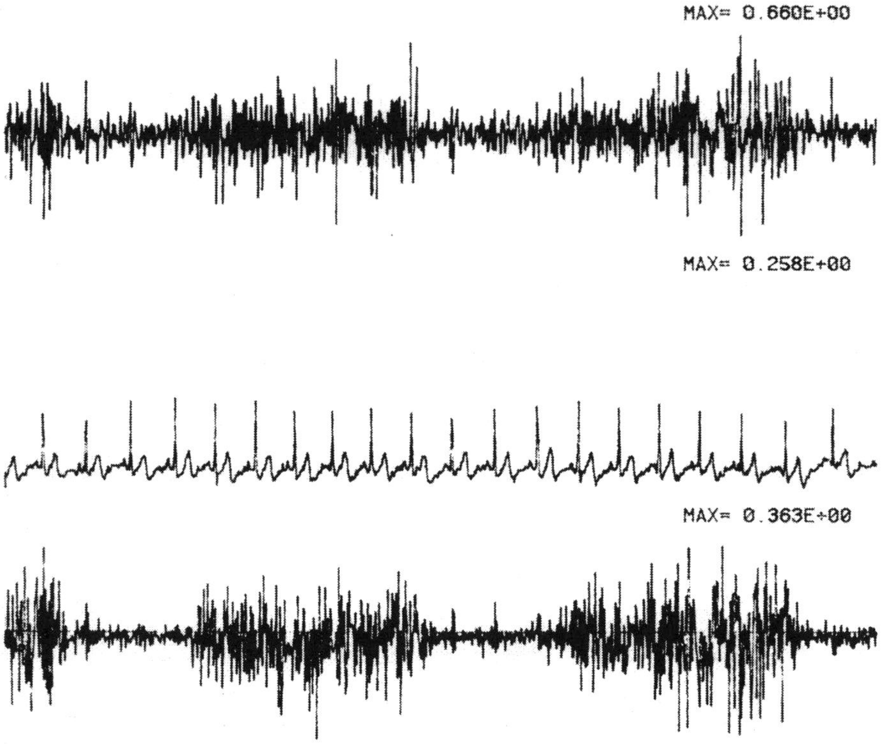

FIGURE 9. EMG and ECG during breathing. Upper trace: EMG from the diaphragmatic muscle; middle: ECG; lower: EMG from the intercostal muscle.

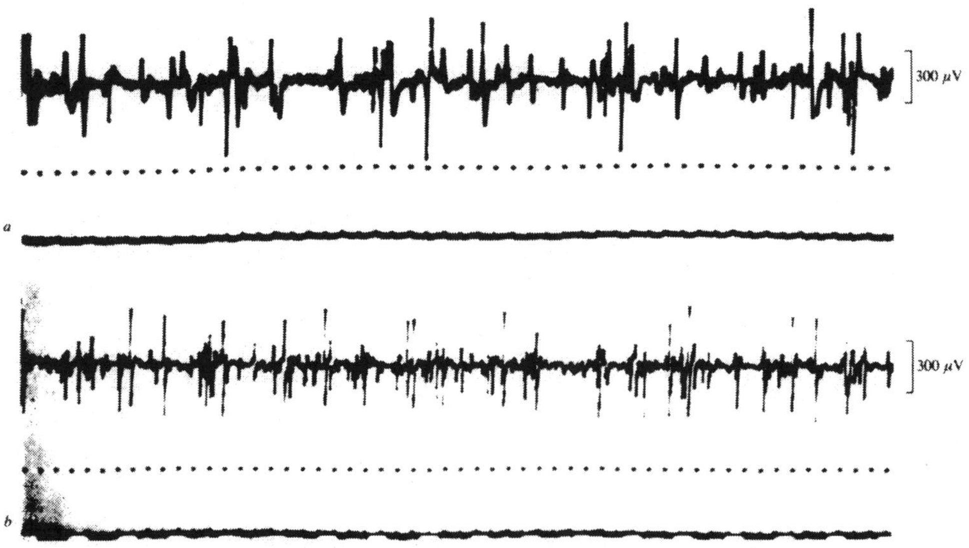

FIGURE 8. Normal motor unit potentials (MUP) recorded from (a) the first dorsal interosseus muscle and (b) the frontalis muscle. (From Lenman, J. A. R. and Ritchie, A. E., *Clinical Electromyography*, Pitman Medical and Scientific, London, 1970. With permission.)

the R-R interval. During sleep the heart rate slows down *(bradycardia)*. It accelerates *(tachycardia)* during exercise, emotional stress, or fever. Rhythm disturbances, arrhythmia, may arise under several abnormal conditions. Sometimes a portion of the myocardium

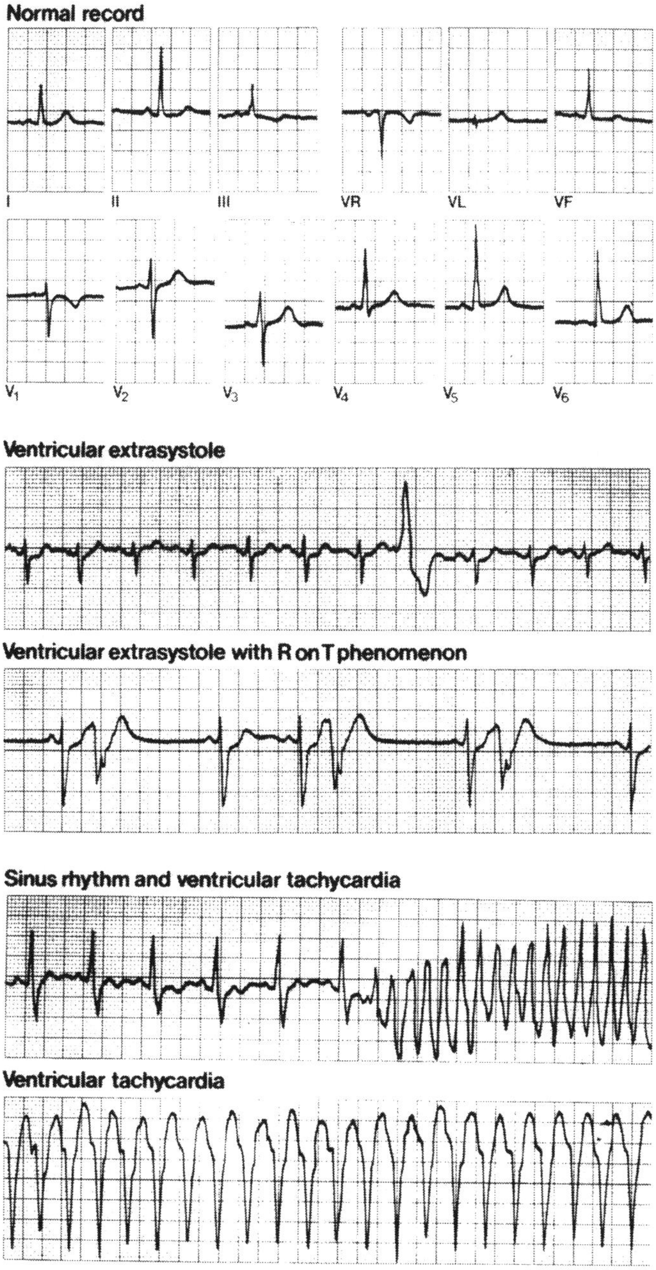

FIGURE 10. Electrocardiograms. (Courtesy of the Pharmaceutical Division, Hoechst, U.K.)

discharges independently, causing a heartbeat not in the normal SA sequence; this is known as *actopic beat extrasystole* or *preventricular contraction* (PVC). When independent discharges continue, the heart may enter a state of *atrial* or *ventricular fibrillation.* Sometimes the cause to these phenomena is a block in the normal neural paths in the heart (i.e., the His bundle or Purkinje fibers). Fibure 10 depicts several clinical ECG recordings.

The processing of the ECG signal is done with the goal of learning about the heart's condition by means of the convenient electrical signal. The *inverse problem*[44,45] in electrocardiography is defined as the problem of determining the electrical sources in the heart at each instant of time, given the electrical potentials at the body surface. This is, however, a difficult task mainly due to the complex volume conductor. Most ECG analysis and diagnosis,[46] however, are performed directly from the surface recordings.

Conventional ECG consists of the PQRST complex with amplitudes of several millivolts. It is usually processed in the frequency band of 0.05 to 100 Hz[47] where most of the energy of the ECG is included.

The first step in ECG processing is the identification of the R wave. This is done in order to synchronize consecutive complexes and for R-R interval (heart rhythm) analysis. Various techniques of wavelet detection have been employed[48] (Chapter 1, Volume II); the problem is particularly severe when recording the ECG under active conditions where muscle signals and other noise sources obscure the QRS complex. The analysis of the R-R interval is an important part of heart patient monitoring. Several methods have been employed for the analysis, among them are autoregressive prediction[49] and state estimation.[50]

Much effort has been placed on the development of algorithms for automatic processing[51,52] of the ECG for monitoring, data compression, and classification. Optimal features[53] of the ECG have been discussed and a variety of methods,[54-57] including linear prediction[54,55] and Karhunen-Loeve expansion,[56] have been employed for compression and classification.

2. High-Frequency Electrocardiography

It has been found that the higher-frequency band of 100 to 1000 Hz filtered out in the normal ECG does contain additional information.[58-60] Waveforms known as *notches* and *slurs* which are superimposed on the slowly varying QRS complexes have been recorded.

3. Fetal Electrocardiography (FECG)

The nonivasive detection of fetal ECG by means of abdominal surface electrodes[61] is used in clinical practice. The main problem in the processing of FECG is the large interferences from maternal ECG (MECG) and from other muscles. Adaptive filtering methods (Chapter 9, Volume I) have been successfully used for signal to noise enhancement. Other methods[62] have also been suggested.

4. His Bundle Electrography (HBE)

The recording of the electrical field, generated by the His and Purkinje activities,[63,64] is known as the *His bundle electrogram* (HBE). The signal has an amplitude range of about 1 to 10 μV. This low-amplitude range requires synchronized averaging techniques for processing.

5. Vector Electrocardiography (VCG)

Rather then displaying the voltages of the surface electrodes as a function of time, one can plot the voltages of one electrode as a function of another or as a function of some combination of other electrodes. With suitable location of the electrodes, one can thus get the approximate projections of the heart dipole on the transverse plane (x,z), the sagittal plane (y,z), and the frontal plane (x,y). Common electrode combinations[17] are, for example, the *Frank electrode system*, the *axial*, the *tetrahedron*, and the *cube vectorcardiograms*. The vectorcardiogram is given by three two-dimensional plots (the projections on the planes) in which the time is a parameter.

I. Electrogastrography (EGG)

The stomach, like the heart, possesses a pacemaker that generates a sequence of electrical potentials. Unlike the heart, defined pacemaker cells have not been found in the stomach. The cyclic electrical potentials are transmitted through the smooth muscle fibers, causing a slow rhythmic (of the order of 0.05 Hz) mechanical motion. This motion is responsible for mixing, grinding, and propelling the absorbed food.

Electrical potential changes generated by the stomach can be picked up[65] by means of

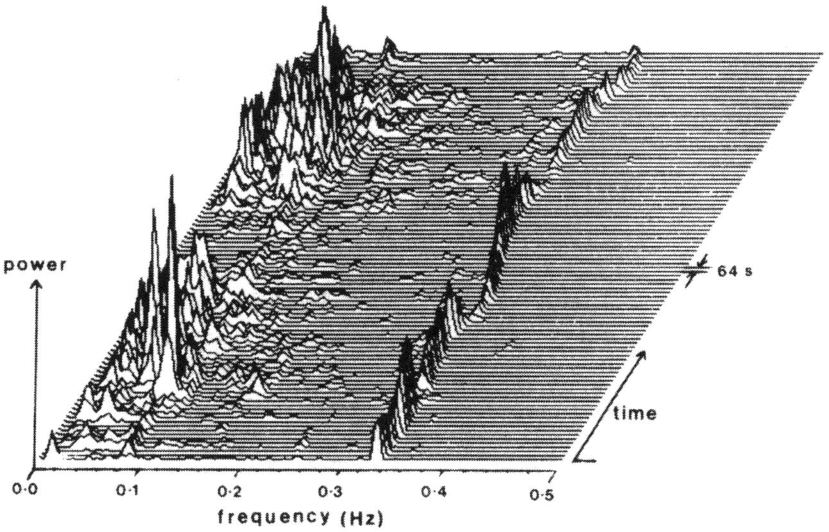

FIGURE 11. Power spectral density functions of dog's electrogastrogram (EGG). Calculated from a record of 107.7 min. The frequency at about 0.32 Hz is of duodenal origin. (From van der Schee, E. J., Electrogastrography Signal Analytical Aspects and Interpretation, Doctoral thesis, University of Rotterdam, The Netherlands, 1984. With permission.)

surface electrodes. The signal has a dominant frequency equal to the frequency of the gastric *electric control activity* (ECA) which is about 0.05 Hz in man.[65] The frequency bandwidth of the signal is about 0.01 to 0.5 Hz. Optimal locations of electrodes for best recordings have been suggested.[66] Interferences due to electrode skin interface and motion and breathing artifacts require signal to noise enhancement techniques. Correlation[67] and adaptive filtering[68] methods have been suggested. Autoregressive analysis of the signal[69] has been used. Using duodenal implanted electrodes, the automatic classification of the ingestion of three different test meals was successfully demonstrated.[70] Pattern recognition methods discussed in Chapter 3 were employed. Figure 11 shows an example of EGG power spectral density function.

J. Galvanic Skin Reflex (GSR), Electrodermal Response (EDR)

The autonomic nervous system, in response to emotional stimulus, changes the activity of the sweat glands. A potential can be detected[17,71] by means of a pair of electrodes: one placed on the palm of the hand (high concentration of glands) and the other on the back of the hand (region almost completely devoid of sweat glands). The signal is less than 1 mV in amplitude with a bandwidth of DC to 5 Hz. The acquisition of the signal is difficult due to DC potentials produced by the electrode skin interface. The GSR is used for emotional state monitoring ("lie detector") and for various biofeedback applications.

III. IMPEDANCE

A. Bioimpedance

The biological tissue obeys Ohm's law for current densities[73] below about 1 mA/cm.^{2} The impedance of the tissue changes with time due to various phenomena such as blood volume change, blood distribution change, blood impedance change with velocity, and tissue impedance changes due to pressure, endocrine, or autonomic nervous system activity. Important information on the resistance of various tissues[74] has been collected through the years.

Bioimpedance measurements[75] are usually performed with four electrodes: two for current injection and two for the impedance measurement. At low frequencies, electrode polarization causes some measurement problems. The range of 50 kHz to 1 MHz is usually employed.

Current densities must be kept low so as not to cause changes due to heating. The range of currents used in practice is 20 μA to 20 mA.

B. Impedance Plethysmography

The use of impedance changes for the recording of peripheral volume pulses is known as *impedance plethysmography*. The method has been applied[75] to various locations of the body such as the digits, limbs, head, thorax, and kidney. Since calibration of the impedance in terms of blood flow is difficult, the method has been mainly used for relative monitoring.

An experiment on a dog at 50 kHz showed that a 1% change in blood volume generates a change of about 0.16% in resistance, with almost linear relationship at a range of ±30% blood flow change.

C. Rheoencephalography (REG)

The measurement of impedance changes between electrodes placed on the scalp is known as *rheoencephalogram* (REG). The frequencies used are in the range of 1 to 500 KHz, yielding a transcarnial impedance of about 100 Ω. The pulsatile impedance change is on the order of 0.1 Ω.

D. Impedance Pneumography

Electrodes placed on the surface of the chest are used to monitor respiration in the frequency range of 50 to 600 kHz; the change in transthoracic impedance, from full inspiration to maximum expiration, is almost entirely resistive with the value of about 20 Ω. The changes in impedance are related to the changes in lung air volumes. The method is used also as an apnea monitor, to detect pauses in breathing.

E. Impedance Oculography (ZOG)

The position of the eye can be monitored by the impedance[75] measured between pairs of electrodes located around the eye.

F. Electroglottography

The measurement of the impedance across the neck in known as *electroglottography*.[72,80] Variations in glottis size, as the vocal cords vibrate, cause impedance changes. The method can thus be used to measure the pitch frequency.

IV. ACOUSTICAL SIGNALS

A. Phonocardiography (PCG)

Phonocardiography[77] is the recording of sounds generated by the heart and the great vessels. PCG can be monitored by invasive means with a microphone inserted into the heart or the vessel. It can be monitored noninvasively by placing a microphone on the surface of the body. The latter is a convenient noninvasive method widely used.

The sounds recorded on the surface of the body depend on the source location, intensity, and the acoustical properties of the surrounding tissues. It is important that the location of the microphone be specified. The required bandwidth is about 20 to 1000 Hz.

The normal heart sounds are divided into four. These are depicted in Figure 12 together with the ECG. The first and second heart sounds are the more important ones. A detailed recording is shown in Figure 13.

1. The First Heart Sound

The first heart sound is the result of the onset of left ventricular contraction (1) (Figure 13), onset of right ventricular contraction, mitral closure (2), onset of right ventricular ejection (3), and onset of left ventricular ejection (4). The first heart sound lasts about 100 to 120 msec.

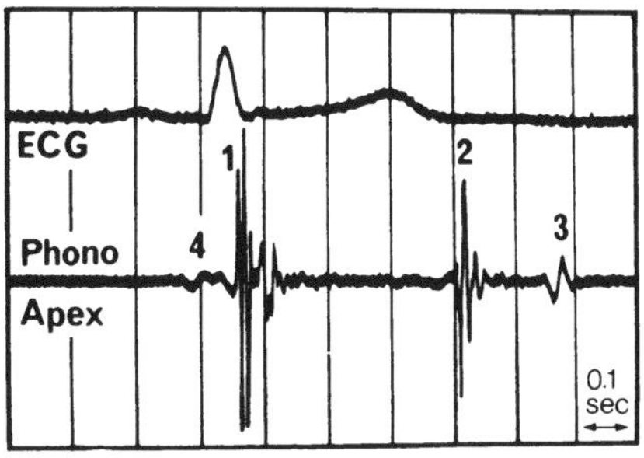

FIGURE 12. (Above) Phonocardiogram, the four heart sounds. (Reproduced with permission from Tavel, M. E., *Clinical Phonocardiography and External Pulse Recording*, 3rd ed., Copyright 1978 by Year Book Medical Publishing, Chicago.)

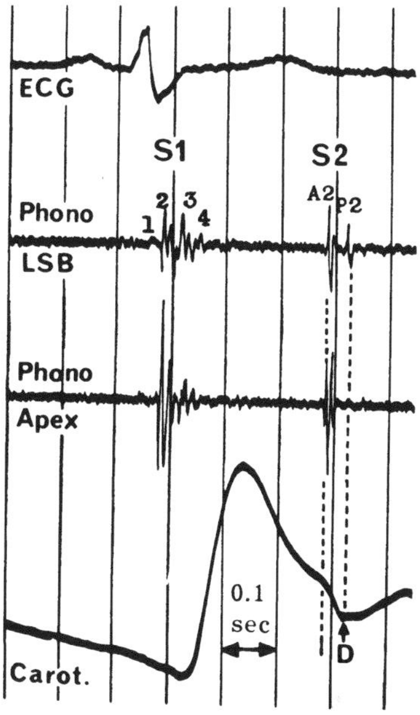

FIGURE 13. (Right) Phonocardiogram, the first and second heart sounds recorded at the left sternal border, fourth intercostal space (LSB), and at the apex. Upper trace, ECG; lower trace, the carotid pulse. Note the dictotic notch (D). (Reproduced with permission from Tavel, M. E., *Clinical Phonocardiography and External Pulse Recording*, 3rd ed., Copyright ©1978 by Year Book Medical Publishing, Chicago.)

2. *The Second Heart Sound*

The second heart sound is generated by the closure of the semilunar valve. Two components can usually be detected (Figure 13) which are the result of the aortic and pulmonary valves.

3. *The Third Heart Sound*

The source of the third heart sound is not agreed upon. It is a low-frequency (20 to 70 Hz) transient with low amplitudes, occuring during ventricular filling of early diastole. The third heart sound lasts about 40 to 50 msec.

4. *The Fourth Heart Sound*

The fourth heart sound occurs at the time of atrial contraction. It is similar to the third heart sound in duration and bandwidth.

5. *Abnormalities of the Heart Sound*

The diagnostic value of the phonocardiography stems from the ability to analyze heart malfunctions from abnormal recordings. Abnormalities appear in many forms; the most important are changes in intensity, splitting of sound components, *ejection clicks* and sounds, *opening snaps,* and *murmurs.* A few examples are shown in Figure 14. Automatic classification systems for PCGs with AR[78] and ARMA[79] analysis have been suggested.

B. Auscultation

The monitoring of sounds heard over the chest walls is known as *auscultation.* It has long been used as one of the means by which pulmonary dysfunctions were diagnosed.[81] During respiration, gases flow through the various airways emitting acoustical energy. This energy is transmitted through the airway walls, the lung tissue, and chest walls.

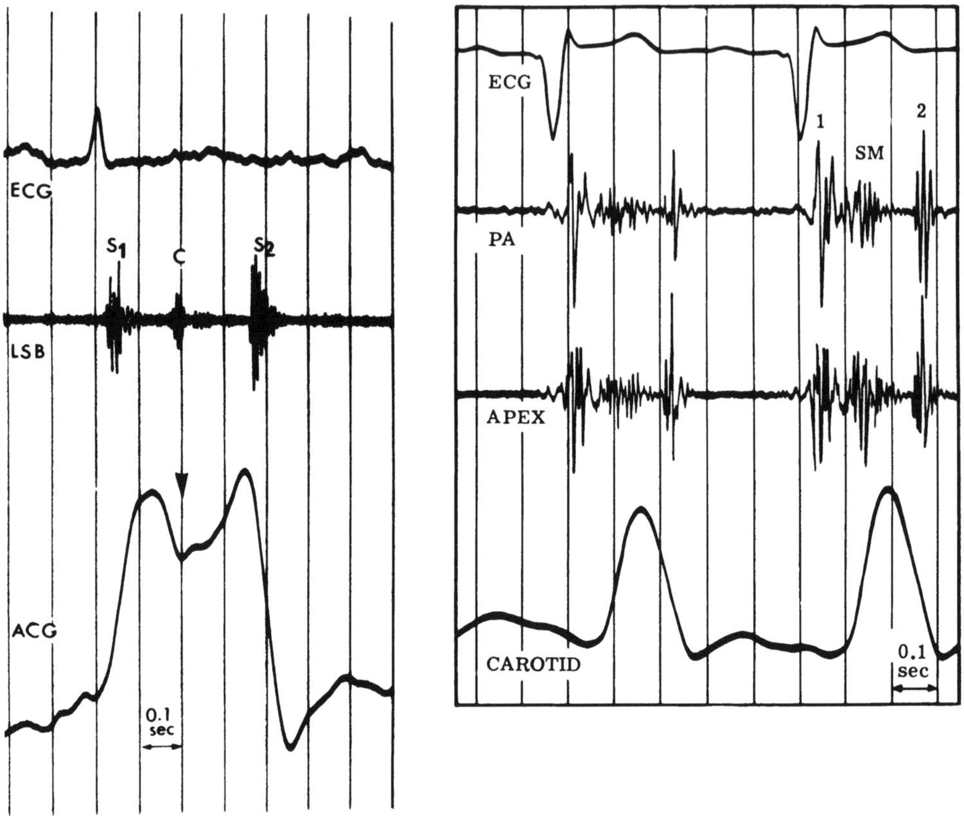

FIGURE 14. Abnormal heart sounds. (A) Midsystolic click; upper trace, ECG; lower trace, apexcardiogram (ACG); (B) systolic ejection murmur. (Reproduced with permission from Tavel, M. E., *Clinical Phonocardiography and External Pulse Recording*, 3rd ed., Copyright ©1978 by Year Book Medical Publishing, Chicago.)

Breath sounds are generated by the air entering the alveoli during inspiration *(local* or *vesicular noise)* and while passing through the larynx *(laryngial* or *glottic hiss)*. Four types of normal breath sounds have been defined: *vesicular breath sounds* (VBS), *bronchial breath sounds* (BBS), *broncho-vesicular breath sounds* (BVBS), and *trachial breath sounds* (TBS). Each one of the above breath sounds is normally heard over certain areas of the thorax. When heard over other than its normal place, it is considered abnormal. Figure 15 depicts the characteristics of the four types of normal breath sounds.

There are several types of breath sounds which, when present, always indicate abnormality. The abnormal breath sounds are known as the *cogwheel breath sound* (CO), the *asmatic breath sound* (AS), the *amphoric breath sound* (AM), and the *cavernous breath sound* (CA). Verbal descriptions of the characteristics of the various breath sounds are used.[81] A parametric description and an automatic classification method have been suggested.[82] Another type of abnormal sounds are the *adventitious* sounds. These are called *musical rales* or *wheezes* and *nonmusical rales*.

Auscultation is usually performed with the stethoscope. To get the full frequency range and an electrical signals that can be processed, microphones are used. The frequency range required is 20 Hz to 2 kHz.

C. Voice

Speech is produced by expelling air from the lungs through the trachea to the vocal cords.[83] When uttering voiced sounds, the vocal cords are forced to be opened by the air pressure. The opening slit is known as the *glottis*. The pulse of air propogates through the vocal tract.

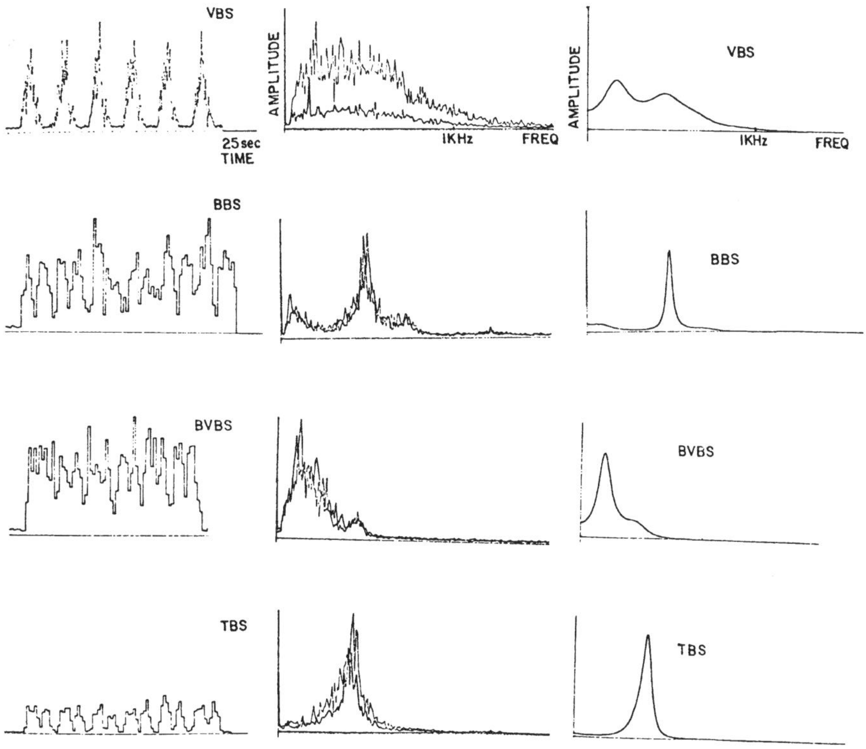

FIGURE 15. Typical time and frequency plots of normal breath sounds. Left: energy envelope; middle: power spectral density, estimated by FFT (upper trace, midinspiration; lower trace, beginning inspiration); right: power density, estimated by LPC (midinspiration). (From Cohen, A. and Landsberg, D., *IEEE Trans. Biol. Med. Eng.*, BME-31, 35, 1984 (© 1984, IEEE). With permission.)

The generated sound depends on the acoustical characteristics of the various tubes and cavities of the vocal system. These are changing during the speech process by moving the tongue, the lips, or the velum.

The frequency of oscillation of the vocal cords during voiced speech is called the *fundamental frequency* or *pitch*. This frequency is determined by the subglottal pressure and by the characteristics of the cords, their elasticity, compliance, mass, length, and thickness.

When uttering unvoiced sounds, the vocal cords are kept open and do not take part in the sound generation. Figure 16 depicts a record of speech signals including silent, unvoiced, and voiced segments. The speech signal has been used as a diagnostic aid for laryngeal pathology or disorder;[80,84,85] among these are laryngitis, hyperplasia, cancer, paralysis, and more. It has been also used as a diagnostic aid for some neurological diseases[86] and as an indicator of emotional states.[87] Infant's cry has also been suggested as a diagnostic aid[88] (e.g., see Figure 2, Chapter 3).

D. Korotkoff Sounds

The most common method for indirect blood pressure measurement is by means of the *sphygmomanometer*. An inflatable cuff placed above the arm is used to oclude blood flow to the arm. The pressure exerted by the cuff causes the artery to collapse. When cuff pressure is gradually released to the point where it is just below arterial pressure, blood starts to flow through the compressed artery. The turbulent blood flow generates sounds known as Korotkoff sounds. These are picked up by a microphone (or a stethoscope) placed over the artery. The sounds continue, while decreasing the pressure, until no constriction is exerted

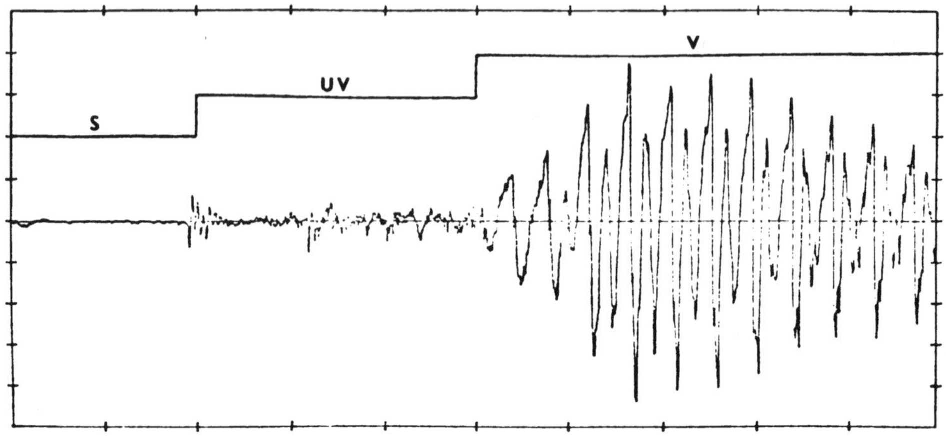

FIGURE 16. A sample of speech signal demonstrating silent, unvoiced, and voiced segments.

on the artery. Most of the sound's power is in the frequency range of 150 to 500 Hz. Usually piezoelectric microphones are used yielding amplitudes of about 100 mV (peak to peak).

V. MECHANICAL SIGNALS

A. Pressure Signals

Blood pressure measurements[13] are taken from the critically ill patient by the insertion of a pressure transducer somewhere in the circulatory system. Figure 13 and Figure 12 in Chapter 4 give typical examples of the carotid blood pressure signal. Pattern recognition methods have been applied to the analysis of the pressure blood wave (Chapter 4). The frequency bandwidth required is about DC to 50 Hz. Other biological pressure signals are of clinical importance. Figure 17, for example, depicts the intrauterine pressure of a woman in labor.

B. Apexcardiography (ACG)

Tavel[77] has suggested the term apexcardiography to include a variety of methods used for recording the movements of the precordium. Among the various methods are *vibrocardiography, kinetocardiography, ballistocardiography,* and *impulse cardiography.* The motion is detected by various transducers, accelerometers, strain gauges, or displacement devices (LVDT). The frequency bandwidth required is about DC to 40 Hz. An example of the ACG is shown in Figure 14A.

C. Pneumotachography

Pneumotachography is a method used to analyze flow rate for respirator functions evaluation. The flow rate signal has a bandwidth of about DC to 40 Hz.

D. Dye and Thermal Dilution

The total amount of blood flowing through the heart is known as the cardiac output. *Dye dilution* methods have been developed to calculate the *cardiac output.* A known amount of dye (or radioisotope) is injected into the superior vena cava. The concentration of dye (isotope) is then monitored in the arteries. By integrating the dilution curve, an estimate of the cardiac output is given. The dilution curve contains the contribution of the blood pumped after injection which is the signal required. Superimposed on it are contributions from the recirculating blood due to the second and third heart cycles. For the cardiac output calcu-

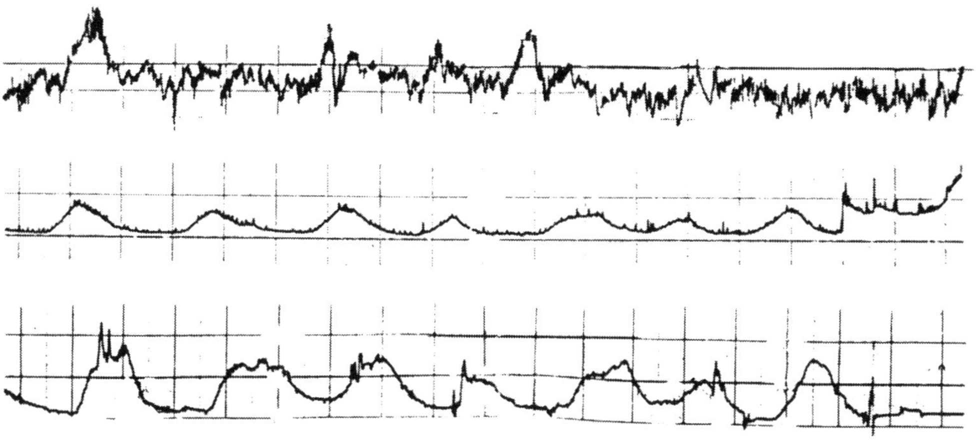

FIGURE 17. Recording during labor. Upper trace: fetal heart rate; middle: abdominal pressure; lower: intrauterine pressure. (Courtesy of Dr. Yarkoni, Soroka Medical Center).

lations, the first curve must be estimated. The techniques for echo cancellation (Chapter 9, Volume I) can be employed here. A similar technique, using the injection of fluid having temperature different than that of the blood, is sometimes employed. It is known as *thermal dilution*.

E. Fetal Movements

The movements of the fetus are thought to give an indication of its well-being. Various methods have been suggested for the noninvasive monitoring of fetal movement. Abdominal strain gauges have been used as well as ultrasonic transducers.

VI. BIOMAGNETIC SIGNALS

A. Magnetoencephalography (MEG)

Various organs such as the heart, lungs, and brain produce extreme weak magnetic fields. The measurement of these magnetic fields is difficult. Magnetic measurement has been made on nerve cells[89] and from the brain.[90] MEG was reported to be different than the EEG and to provide additional information.[90] An example of the MEG signal is shown in Figure 18.

B. Magnetocardiography (MCG)

The magnetic fields generated by the heart were reported to be different from the ECG and to provide additional information.[91] Figure 19 depicts the MCG taken from various places on the chest.

C. Magnetopneumography (MPG)

The monitoring of the magnetic fields generated over the lungs was also suggested.[92]

VII. BIOCHEMICAL SIGNALS

Biochemical measurements[1] are usually performed in the clinical laboratory. Blood gas and acid-base measurements are routinely performed to evaluate partial pressure of oxygen (pO_2), partial pressure of CO_2 (pCO_2), and concentration of hydrogen ions (pH). These measurements are usually done by means of electrodes. Other methods for the measurements of organic and nonorganic chemical substances are used, such as *chromatography*, *electrophoresis*, *flame photometry*, *atomic emission*, and *absorption fluorometry*, *nuclear magnetic*

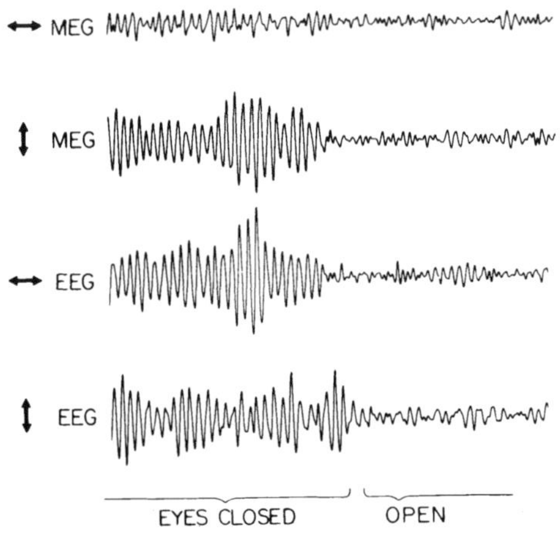

FIGURE 18. Magnetoencephalogram (MEG) with EEG. (From Cohen, D. and Cuffin, B. N., *Electroencephalogr. Clin. Neurophys.*, 56, 1983. With permission.)

resonance (NMR), and more. These methods most often provide DC signals. The problems associated are mainly in the instrumentation and acquisition systems rather than in the processing. Some processing problems do exist, for example, in methods like chromatography where sometimes close or overlapping peaks have to be identified.

Biochemical measurements are performed also in the clinic and in the research laboratory. Specific ion microelectrodes have been developed which allow the recordings of ion concentration variations of neural cells. Figure 20 is an example of such a signal.

Noninvasive, transcutaneous monitoring of pO_2 and pCO_2 can be conveniently performed by means of special electrodes. This measurement is used in the clinic. Noninvasive blood oxygenation monitoring is done by optical means (oximetry). These signals are very low-frequency signals and usually require no special processing.

VIII. TWO-DIMENSIONAL SIGNALS

The problem of image processing is an important problem in biomedical diagnosis. Methods that use X-rays and ultrasound provide two-dimensional images of the tissues under investigation. Sophisticated imaging systems like the *computer tomography* (CT) and *nuclear magnetic resonance* (NMR) scanners reconstruct three-dimensional information from the two-dimensional measurements. The analysis and processing of these types of signals are outside the scope of this book.

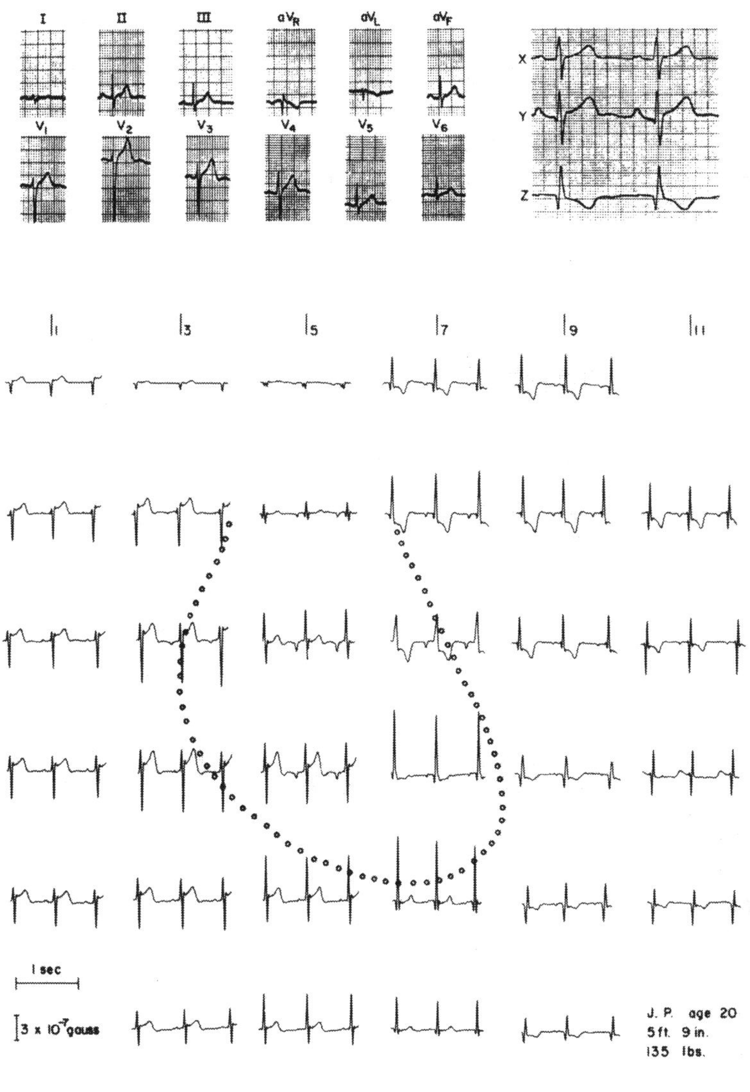

FIGURE 19. Magnetocardiogram obtained across the chest with 12-lead ECG and Frank x,y,z leads. (From Cohen, D. and McCaughan, D., *Am. J. Cardiol.*, 29, 678, 1972. With permission.)

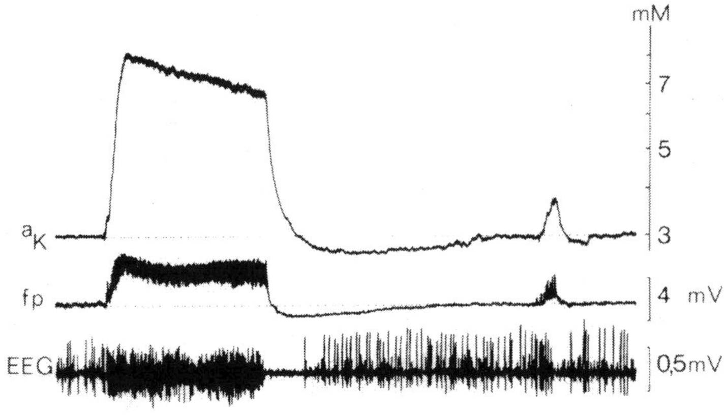

FIGURE 20. Simultaneous recordings of EEG, field potentials (f_p), and extracellular potassium activity (a_k) from cat thalamus during penicillin-induced seizures. (From Heinemann, U. and Gutnick, M. J., *Electroencephalogr. Clin. Neurophys.*, 47, 345, 1979. With permission.)

REFERENCES

1. **Webster, J. G., Ed.,** *Medical Instrumentation Application and Design*, Houghton Mifflin, Boston, 1978.
2. **Abeles, M. and Goldstein, M. H.,** Multispike train analysis, *Proc. IEEE*, 65(5), 762, 1977.
3. **Lenman, J. A. R. and Ritchie, A. E.,** *Clinical Electromyography*, Pitman Medical and Scientific, London, 1970.
4. **Armington, J.,** *The Electroretinogram*, Academic Press, New York, 1974.
5. **Gouras, P.,** Electroretinography: some basic principles, *Invest. Ophthalmol.*, 9, 557, 1970.
6. **Chatrian, G. E.,** Computer assist ERG. I. Standardized method, *Am. J. EEG Technol.*, 20(2), 57, 1980.
7. **Larkin, R. M., Klein, S., Odgen, T. E., and Fender, D. H.,** Non-linear kernels of the human ERG, *Biol. Cybern.*, 35, 145, 1979.
8. **Krill, A. E.,** The electroretinogram and electro-oculogram: clinical applications, *Invest. Ophthalmol.*, 9, 600, 1970.
9. **North, A. W.,** Accuracy and precision of electro-oculographic recordings, *Invest. Ophthalmol.*, 4, 343, 1965.
10. **Kris, C.,** Vision: electro-oculography, in *Medical Physics*, Vol. 3, Glasser, O., Ed., Year Book Medical Publishing, Chicago, 1960.
11. **Kiloh, L. G., McComas, A. J., Osselton, J. W., and Upton, A. R. M.,** *Clinical Electroencephalography*, 4th ed., Butterworths, London, 1981.
12. **Basar, E.,** *EEG-Brain Dynamics*, Elsevier/North-Holland, Amsterdam, 1980.
13. **Cox, J. R., Nolle, F. M., and Arthur, R. M.,** Digital analysis of the EEG, the blood pressure and the ECG, *Proc. IEEE*, 60, 1137, 1972.
14. **Barlow, J. S.,** Computerized clinical EEG in perspective, *IEEE Trans. Biol. Med. Eng.*, 26, 277, 1979.
15. **Gevins, A. S.,** Pattern recognition of human brain electrical potentials, *IEEE Trans. Pattern Anal. Mach. Intelligence*, 2, 383, 1980.
16. **Isaksson, A., Wennberg, A., and Zetterberg, L. H.,** Computer analysis of EEG signals with parametric models, *Proc. IEEE*, 69, 451, 1981.
17. **Strong, P.,** *Biophysical Measurements*, Tektronix, Beaverton, Ore., 1970.
18. **Childers, D. G.,** Evoked responses: electrogenesis, models, methodology and wavefront reconstruction and tracking analysis, *Proc. IEEE*, 65(5), 611, 1977.
19. **McGillem, D. C. and Aunon, J. I.,** Measurements of signal components in single visually evoked brain potentials, *IEEE Trans. Biol. Med. Eng.*, 24, 232, 1977.
20. **Sayers, B. McA., Beagley, H. A., and Riha, J.,** Pattern analysis of auditory evoked EEG potentials, *Audiology*, 18, 1, 1979.
21. **Jervis, B. W., Nichols, M. J., Johnson, T. E., Allen, E., and Hudson, N. R.,** A fundamental investigation of the composition of auditory evoked potentials, *IEEE Trans. Biol. Med. Eng.*, 30, 43, 1983.

22. **Boston, J. R.,** Spectra of auditory brainstem responses and spontaneous EEG, *IEEE Trans. Biol. Med. Eng.*, 28, 334, 1981.
23. **Sclabussi, R. J., Risch, H. A., Hinman, C. L., Kroin, J. S., Enns, N. F., and Namerow, N. S.,** Complex pattern evoked somatosensory responses in the study of multiple sclerosis, *Proc. IEEE*, 65(5), 626, 1977.
24. **Berger, M. D.,** Analysis of sensory evoked potentials using normalized cross-correlation functions, *Med. Biol. Eng. Comput.*, 21, 149, 1983.
25. **Carmon, A.,** Consideration of the cerebral response to painful stimulation: stimulus transduction versus perceptual event, *Bull. N.Y. Acad. Med.*, 55, 313, 1979.
26. **Zetterberg, L. H.,** Estimation of parameters for a linear difference equation with application to EEG analysis, *Math. Biosci.*, 5, 227, 1969.
27. **Praetorius, H. M., Bodenstein, G., and Creutzfeldt, O. D.,** Adaptive segmentation of EEG records, a new approach to automatic EEG analysis, *Electroencephalogr. Clin. Neurophys.*, 42, 84, 1977.
28. **Haimi-Cohen, R. and Cohen, A.,** A microprocessor controlled system for stimulation and acquisition of evoked potentials, *Comput. Biomed. Res.*, in press.
29. **Basmajian, J., Clifford, H., McLeod, W., and Nunnaly, H., Eds.,** *Computers in Electromyography*, Butterworths, London, 1975.
30. **Stalberg, E. and Antoni, L.,** Computer aided EMG analysis, in *Computer Aided Electromyography, Progress in Clinical Electromyography*, Vol. 10, Desmedt, J. E., Ed., S. Karger, Basel, 1983, 186.
31. **LeFever, R. S. and DeLuca, C. J.,** A procedure for decomposing the myoelectric signal into its constituent action potentials. I. Techniques, theory and implementation, *IEEE Trans. Biol. Eng.*, 29, 149, 1982.
32. **LeFever, R. S., Xenakis, A. P., and DeLuca, C. J.,** A procedure for decomposing the myoelectric signal into its constituent action potentials. II. Execution and test for accuracy, *IEEE Trans. Biol. Med. Eng.*, 29, 158, 1982.
33. **Nandedkar, S. D. and Sanders, D. B.,** Special purpose orthonormal basis functions — application to motor unit action potentials, *IEEE Trans. Biol. Med. Eng.*, 31, 374, 1984.
34. **Berzuini, C., Maranzana-Figini, M., and Bernardinelli, C.,** Effective use of EMG parameters in the assessment of neuromuscular diseases, *Int. J. Biol. Med. Comput.*, 13, 481, 1982.
35. **Kranz, H., Williams, A. M., Cassell, J., Caddy, D. J., and Silberstein, R. B.,** Factors determining the frequency content of the EMG, *J. Appl. Physiol. Respir. Environ. Exerc. Physiol.*, 55(2), 392, 1983.
36. **Lindstrom, L. H. and Magnusson, R. I.,** Interpretation of myoelectric power spectra: a model and its application, *Proc. IEEE*, 65, 653, 1977.
37. **Inbar, G. F. and Noujaim, A. E.,** On surface EMG spectral characterization and its application to diagnostic classifications, *IEEE Trans. Biol. Med. Eng.*, 31, 597, 1984.
38. **Journee, J. L., van-Manen, J., and van-der Meer, J. J.,** Demodulation of EMG's of pathological tremours. Development and testing of a demodulator for clinical use, *Med. Biol. Eng. Comput.*, 21, 172, 1983.
39. **Stulen, F. G. and DeLuca, C. J.,** Muscle fatigue monitor: a non-invasive device for observing localized muscular fatigue, *IEEE Trans. Biol. Eng.*, 29, 760, 1982.
40. **Gross, D., Grassino, A., Ross, W. R. D., and Macklem, P. T.,** Electromyogram pattern of diaphragmatic fatigue, *J. Appl. Physiol. Respir. Environ. Exerc. Physiol.*, 46(1), 1, 1979.
41. **Graupe, D. and Cline, W. K.,** Functional separation of EMG signals via ARMA identification methods for prosthesis control purposes, *IEEE Trans. Syst. Man Cybern.*, 5, 252, 1975.
42. **Doerschuk, P. C., Gustafson, D. E., and Willsky, A. S.,** Upper extremity limb function discrimination using EMG signal analysis, *IEEE Trans. Biol. Med. Eng.*, 30, 18, 1983.
43. **Saridis, G. N. and Gootee, P. T.,** EMG pattern analysis and classification for a prosthetic arm, *IEEE Trans. Biol. Med. Eng.*, 29, 403, 1982.
44. **Martin, R. O., Pilkington, T. C., and Marrow, M. N.,** Statistically constrained inverse electrocardiography, *IEEE Trans. Biol. Med. Eng.*, 22, 487, 1975.
45. **Yamashita, Y.,** Theoretical studies on the inverse problem in electrocardiography and the uniqueness of the solution, *IEEE Trans. Biol. Med. Eng.*, 29, 719, 1982.
46. **Friedman, H. H.,** *Diagnostic Electrocardiography and Vectorcardiography*, McGraw-Hill, New York, 1977.
47. **Riggs, T., Isenstein, B., and Thomas, C.,** Spectral analysis of the normal ECG in children and adults, *J. Electrocardiol.*, 12(4), 377, 1979.
48. **Ligtenberg, A. and Kunt, M.,** A robust digital QRS detection algorithm for arrhythmia monitoring, *Comput. Biomed. Res.*, 16, 273, 1983.
49. **Haywood, L. Y., Saltzberg, S. A., Murthy, V. K., Huss, R., Harvey, G. A., and Kalaba, R.,** Clinical use of R-R interval prediction for ECG monitoring: time series analysis by autoregressive models, *Med. Inst.*, 6, 111, 1972.
50. **Ciocloda, G. H.,** Digital analysis of the R-R intervals for identification of cardiac arrhythmia, *Int. J. Biol. Med. Comput.*, 14, 155, 1983.

51. **Caceres, C. A. and Dreifus, L. S., Eds.,** *Clinical Electrocardiography and Computers,* Academic Press, New York, 1970.
52. **Wolf, H. K. and MacFarlane, P. W., Eds.,** *Optimization of Computer ECG Processing,* North-Holland, Amsterdam, 1980.
53. **Jain, U., Rautaharju, P. M., and Warren, J.,** Selection of optimal features for classification of electrocardiograms, *J. Electrocardiol.,* 14(3), 239, 1981.
54. **Shridhar, M. and Stevens, M. F.,** Analysis of ECG data, for data compression, *Int. J. Biol. Med. Comput.,* 10, 113, 1979.
55. **Ruttimann, U. E. and Pipberger, H. V.,** Compression of the ECG by prediction or interpolation and entropy encoding, *IEEE Trans. Biol. Med. Eng.,* 26, 163, 1979.
56. **Womble, M. E., Halliday, J. S., Mitter, S. K., Lancaster, M. C., and Triebwasser, J. H.,** Data compression for storing and transmitting ECG's/VCG's, *Proc. IEEE,* 65(5), 702, 1977.
57. **Jain, U., Rautaharju, P. M., and Horacek, B. M.,** The stability of decision theoretic electrocardiographic classifiers based on the use of discretized features, *Comput. Biomed. Res.,* 13, 132, 1980.
58. **Santopietro, R. F.,** The origin and characteristics of the primary signal, noise and interference sources in the high frequency ECG, *Proc. IEEE,* 65(5), 707, 1977.
59. **Chein, I. C., Tompkins, W. J., and Britler, S. A.,** Computer methods for analysing the high frequency ECG, *Med. Biol. Eng. Comput.,* 18, 303, 1980.
60. **Kim, Y. and Tompkins, W. J.,** Forward and inverse high frequency ECG, *Med. Biol. Eng. Comput.,* 19, 11, 1981.
61. **Wheeler, T., Murrills, A., and Shelly, T.,** Measurement of the fetal heart rate during pregnancy by a new electrocardiography technique, *Br. J. Obstet. Gynaecol.,* 85, 12, 1978.
62. **Bergveld, P. and Meijer, W. J. H.,** A new technique for the suppression of the MECG, *IEEE Biol. Med. Eng.,* 28, 348, 1981.
63. **Flowers, N. C., Hand, R. C., Orander, P. C., Miller, C. R., and Walden, M. O.,** Surface recording of electrical activity from the region of the bundle of His, *Am. J. Cardiol.,* 33, 384, 1974.
64. **Peper, A., Jonges, R., Losekoot, T. G., and Grimbergen, C.,** Separation of His-Purkinge potentials from coinciding atrium signals: removal of the P-wave from the electrocardiogram, *Med. Biol. Eng. Comput.,* 20, 195, 1982.
65. **van der Schee, E. J.,** Electrogastrography Signal Analytical Aspects and Interpretation, Doctoral thesis, University of Rotterdam, The Netherlands, 1984.
66. **Mirizzi, N. and Scafoglieri, U.,** Optimal direction of the EGG signal in man, *Med. Biol. Eng. Comput.,* 21, 385, 1983.
67. **Postaire, J. G., van Houtte, N., and Devroede, G.,** A computer system for quantitative analysis of gastrointestinal signals, *Comput. Biol. Med.,* 9, 295, 1979.
68. **Kentie, M. A., van der Schee, E. J., Grashuis, J. L., and Smout, A. J. P. M.,** Adaptive filtering of canine EGG signals. II. Filter performance, *Med. Biol. Eng. Comput.,* 19, 765, 1981.
69. **Kwok, H. H. L.,** Autoregressive analysis applied to surface and serosal measurements of the human stomach, *IEEE Trans. Biol. Med. Eng.,* 26, 405, 1979.
70. **Reddy, S. N., Dumpala, S. R., Sarna, S. K., and Northeott, P. G.,** Pattern recognition of canine duodenal contractile activity, *IEEE Trans. Biol. Med. Eng.,* 28, 696, 1981.
71. **Vdow, M. R., Erwin, C. W., and Cipolat, A. L.,** Biofeedback control of skin potential level, *Biofeedback Self Regul.,* 4(2), 133, 1979.
72. **Askenfeld, A.,** A comparison of contact microphone and electroglottograph for the measurement of vocal fundamental frequency, *J. Speech Hearing Res.,* 23, 258, 1980.
73. **Schwan, H. P.,** Alternating current spectroscopy of biological substances, *Proc. IRE,* 47(11), 1941, 1959.
74. **Geddes, L. A. and Baker, L. E.,** The specific resistance of biological material. A compendium of data for the biomedical engineer and physiologist, *Med. Biol. Eng. Comput.,* 5, 271, 1967.
75. **Geddes, L. A. and Baker, L. E.,** *Principles of Applied Biomedical Instrumentation,* John Wiley & Sons, New York, 1968.
76. **Lifshitz, K.,** Electrical impedance cephalography (rheoencephalography), in *Biomedical Engineering Systems,* Clynes, M. and Milsum, J. H., Eds., McGraw-Hill, New York, 1970.
77. **Tavel, M. E.,** *Clinical Phonocardiography and External Pulse Recording,* 3rd ed., Year Book Medical Publishing, Chicago, 1967.
78. **Iwata, A., Suzumura, N., and Ikegaya, K.,** Pattern classification of the phonocardiogram using linear prediction analysis, *Med. Biol. Eng. Comput.,* 15, 407, 1977.
79. **Joo, T. H., McClellan, J. H., Foale, R. A., Myers, G. S., and Lees, R. S.,** Pole-zero modeling and classification of PCG, *IEEE Trans. Biol. Med. Eng.,* 30, 110, 1983.
80. **Childers, D. G.,** Laryngial pathology detection, *CRC Crit. Rev. Bioeng.,* 2, 375, 1977.
81. **Druger, G.,** *The Chest: Its Signs and Sounds,* Humetrics Corp., Los Angeles, 1973.
82. **Cohen, A. and Landsberg, D.,** Analysis and automatic classification of breath sounds, *IEEE Trans. Biol. Med. Eng.,* 31, 35, 1984.

83. **Schafer, R. W. and Markel, J. D., Eds.**, *Speech Analysis*, IEEE Press, New York, 1978.
84. **Mezzalama, M., Prinetto, P., and Morra, B.**, Experiments in automatic classification of laryngeal pathology, *Med. Biol. Eng. Comput.*, 21, 603, 1983.
85. **Deller, J. R. and Anderson, D. J.**, Automatic classification of laryngeal dysfunction using the roots of the digital inverse filter, *IEEE Trans. Biol. Med. Eng.*, 27, 714, 1980.
86. **Okada, M.**, Measurement of speech patterns in neurological disease, *Med. Biol. Eng. Comput.*, 21, 145, 1983.
87. **Streeter, L. A., Macdonald, N. H., Apple, W., Krauss, R. M., and Galott, K. M.**, Acoustic and perceptual indicators of emotional stress, *J. Acoust. Soc. Am.*, 73(4), 1354, 1983.
88. **Cohen, A. and Zmora, E.**, Automatic classification of infants' hunger and pain cry, in *Proc. Int. Conf. Digital Signal Process.*, Cappelini, V. and Constantinides, A. G., Eds., Elsevier, Amsterdam, 1984.
89. **Wikswo, J. P., Barach, J. P., and Freeman, J. A.**, Magnetic field of a nerve impulse: first measurements, *Science*, 208, 53, 1980.
90. **Cohen, D. and Cuffin, B. N.**, Demonstration of useful differences between magnetoencephalogram and electroencephalogram, *Electroencephalogr. Clin. Neurophys.*, 56, 1983.
91. **Cohen, D. and McCaughan, D.**, Magnetocardiograms and their variation over the chest in normal subjects, *Am. J. Cardiol.*, 29, 678, 1972.
92. **Robinson, S. E.**, Magnetopneumography non-invasive imaging of magnetic particulate in the lung and other organs, *IEEE Trans. Nucl. Sci.*, 28, 171, 1981.
93. **Heinemann, U. and Gutnick, M. J.**, Relation between extracellular potassium concentration and neuronal activities in cat thalamus (VPL) during projection of cortical epileptiform discharge, *Electroencephalogr. Clin. Neurophys.*, 47, 345, 1979.

Appendix B

DATA AND LAG WINDOWS

I. INTRODUCTION

Any practical signal processing problem requires the use of a window. Since we cannot process an infinitely long record, we must multiply it with a window that zeroes the signal outside the observation period. The topics of window design and window applications are dealt with by most signal processing books[1-8] and by many papers.[9-20]

A window,[1] w(t), is a real and even function of time, with Fourier transform, W(w) = F{w(t)}, which is also real and even. We also require that a window be normalized:

$$w(o) = \frac{1}{2\Pi}\int_{-\infty}^{\infty} W(w)dw = 1 \qquad (B.1)$$

and time limited:

$$w(t) = 0$$
$$|t| > T \qquad (B.2)$$

Windows are used for a variety of applications in continuous and discrete signal processing, e.g., in the design[6] of the nonrecursive digital filters, in the application of FFT, and in power spectral density (PSD) function estimation (Chapter 8, Volume I).

In the application of PSD estimation, a window is required to reduce the spectral leakage. Several figures of merits have been defined to evaluate and compare windows. To cancel leakage completely we need a window that behaves as a delta function in the frequency domain. Such a window is of course unrealizable. We can consider a practical window (Figure 1) and require that the main half width[10] (MHW) of the main lobe and that the side lobe level[10] (SLL) be as small as possible. Other criteria such as the equivalent noise bandwidth[9] (ENBW), processing gain[9] (PG), maximum energy concentration,[1] and minimum amplitude moment[1] have been used.

When considering PSD estimation, a window can be applied directly to the data (data window or taper window) or to the autocorrelation function. The latter is known as the lag window or quadratic window. Note that the data window does not preserve the energy of the signal. The lag window, however, does preserve the energy since r(o), the signal's energy, is multiplied by w(o) = 1.

II. SOME CLASSICAL WINDOWS

A. Introduction

In this section we shall list a number of windows with their appropriate parameters. Plots of the windows in the time and frequency domain are also given. To demonstrate the relative behavior of the windows in PSD estimation application a simple experiment was conducted by Harris.[9] A signal was synthesized, composed of two sinusoids, one with frequency of 10.5 fs/N and amplitude 1.00 and the other with frequency 16.0 fs/N and amplitude 0.01 (40.0 dB below the first), with N being the number of samples in the window. The PSD

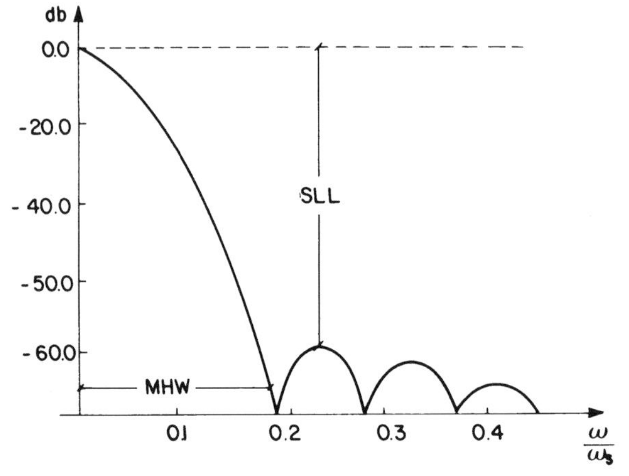

FIGURE 1. A typical window in the frequency domain.

of the signal was estimated by means of the DFT and the various windows. The data given here are from Harris.[9,17] We shall consider here the discrete case only.

The windows are given in terms of even sequences, symmetric with respect to the origin. The sequences (including the sample at the origin) contain odd number of samples. Hence the DFT is complex. The coordinates are normalized with a sampling period, $T_s = 1$. The DFT spans the range of frequencies 0 to 2 Π with frequency difference between two DFT consecutive samples of 2 Π/N.

B. Rectangular (Dirichlet) Window

The most simple window has unity again over the observation record. It is the most desirable window from the point of view of calculations load; however, its performance is usually not satisfactory. The window is given by:

$$w(n) = 1.0$$

$$n = -\frac{N}{2}, \ldots, -1, 0, 1, \ldots, \frac{N}{2} \tag{B.3}$$

and in the frequency domain:

$$W(k) = \exp\left(-j\frac{N-1}{2}k\right)\frac{\text{Sin}\left(\frac{N}{2}k\right)}{\text{Sin}\left(\frac{1}{2}k\right)} \tag{B.4}$$

Figure 2A depicts the window and Figure 2B depicts the estimation of the PSD of the synthesized two-sines signal. Note that the leakage has completely masked the contribution of the second sine wave since the side lobe of the window is much higher than the 40 dB of the sine wave.

C. Triangle (Bartlet) Window

The triangle window was used for the estimation of the correlation coefficient for PSD and AR (Chapters 7 and 8, Volume I). The window is given by:

$$w(n) = 1.0 - \frac{2|n|}{N}$$

$$n = -\frac{N}{2}, \ldots, -1,0,1,\ldots, \frac{N}{2} \qquad \text{(B.5A)}$$

For use with DFT, the window is given by:

$$w(n) = \begin{cases} \dfrac{2n}{N} & ; \; n = 0,1,\ldots, \dfrac{N}{2} \\ 2\left(1 - \dfrac{n}{N}\right) & ; \; n = \dfrac{N}{2}, \ldots, (N - 1) \end{cases} \qquad \text{(B.5B)}$$

and in the frequency domain:

$$W(k) = \frac{2}{N} \exp\left(-j\left(\frac{N}{2} - 1\right)k\right) \frac{\text{Sin}^2\left(\frac{N}{4}k\right)}{\text{Sin}^2\left(\frac{1}{2}k\right)} \qquad \text{(B.6)}$$

Figure 3A depicts the window and Figure 3B depicts the PSD estimation. The MHW is twice that of the rectangle. Its SLL is, however, 26 dB, while that of the rectangle was only 13 dB. Note also that the freuqency function is nonnegative. When the triangle window is used in the PSD estimation, the second sine is barely detectable.

D. Cosine$^\alpha$ Windows

A family of windows is given by:

$$w(n) = \text{Cos}^\alpha\left(\frac{n\Pi}{N}\right)$$

$$n = -\frac{N}{2}, \ldots, -1,0,1,\ldots, \frac{N}{2} \qquad \text{(B.7)}$$

where α is the parameter of the family. Common windows are the windows with α = 1,2,3,4. The window with α = 2 is known as the *Hanning window*. Consider the Hanning window:

$$w(n) = 0.5 + 0.5\,\text{Cos}\left[\frac{2n\Pi}{N}\right] \qquad \text{(B.8)}$$

with

$$W(k) = 0.5\,D(k) + 0.25\left[D\left(k - \frac{2\Pi}{N}\right) + D\left(k + \frac{2\Pi}{N}\right)\right]$$

where

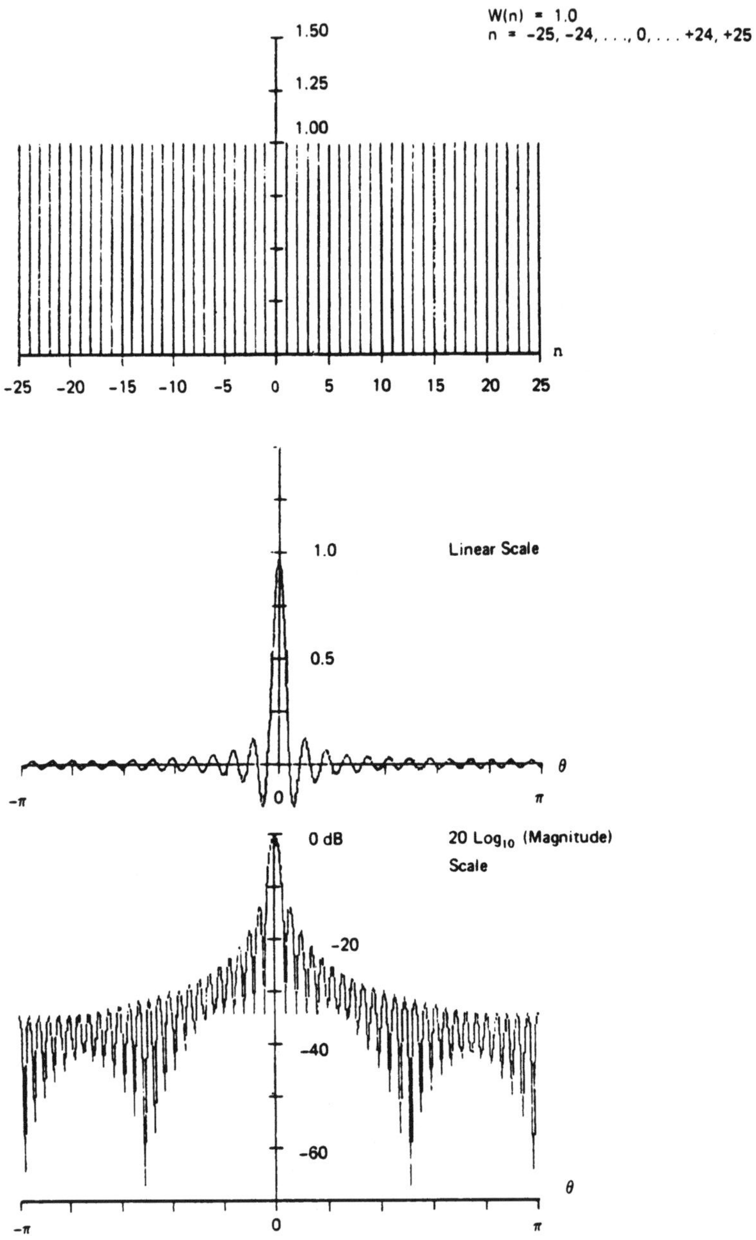

FIGURE 2A. Rectangular (Dirichlet) window in time and frequency domains. Upper trace, the window in the time domain; middle trace, the window in the frequency domain, linear scale; lower trace, the window in the frequency domain, logarithmic scale. (From Harris, F. J., Trigonometric Transforms, A Unique Introduction to the FFT, Tech. Publ. DSP-005 (8-81), Spectral Dynamic Division, Scientific Atlanta, San Diego, 1981. With permission.)

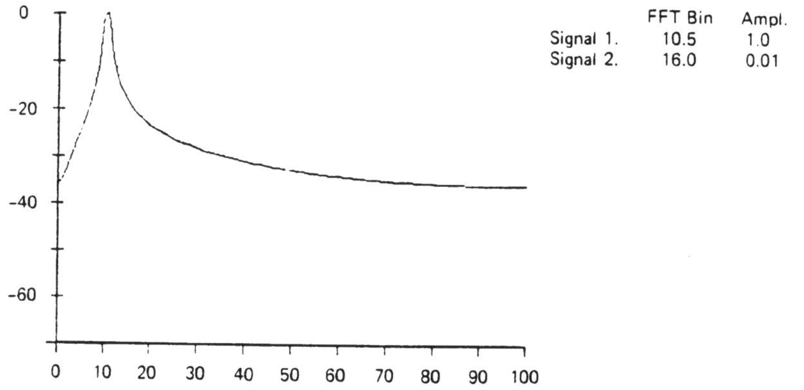

FIGURE 2B. Rectangular (Dirichlet) window in time and frequency domains. FFT power spectral density function estimation of synthesized signal consisting of two sinewaves with frequencies of 10.5 and 16 fs/N and amplitudes of 1.00 and 0.01, respectively. Data window was used (fs, sampling frequency; N, number of samples in the window). (From Harris, F. J., Trigonometric Transforms, A Unique Introduction to the FFT, Tech. Publ. DSP-005 (8-81), Spectral Dynamic Division, Scientific Atlanta, San Diego, 1981. With permission.)

$$D(k) = \exp\left(j\,\frac{k}{2}\right)\frac{\mathrm{Sin}\left(\frac{N}{2}\,k\right)}{\mathrm{Sin}\left(\frac{1}{2}\,k\right)} \tag{B.9}$$

Note that Equation B.9 is given in terms of three Dirichlet kernels, one at the origin and the others shifted one sampling point to the right and to the left. The summation of the three kernels because of phase differences tends to cancel the side lobes.

Figures 4A and B depict the Hanning window, which shows the SLL to be more than 30 dB. The improvement in the detection of the second sine wave through the PSD is clearly seen. The Hanning window has several advantages, one is the fact that the window samples are identical with samples used for the FFT. Hence when FFT is used, the Hanning window requires no additional storage.

E. Hamming Window

The side lobe cancellation in the Hanning is not perfect. It can be improved by adjusting the sizes of the kernels. Optimization of the cancellation yields the *Hamming window*.

$$w(n) = 0.54 + 0.46\ \mathrm{Cos}\left(\frac{2\Pi}{N}\,n\right)$$

$$n = -\frac{N}{2},\ldots, -1,0,1,\ldots,\frac{N}{2} \tag{B.10A}$$

for DFT use,

$$w(n) = 0.54 - 0.46\ \mathrm{Cos}\left(\frac{2\Pi}{N}\,n\right)$$

$$n = 0,1,\ldots,(N-1) \tag{B.10B}$$

with the transformation:

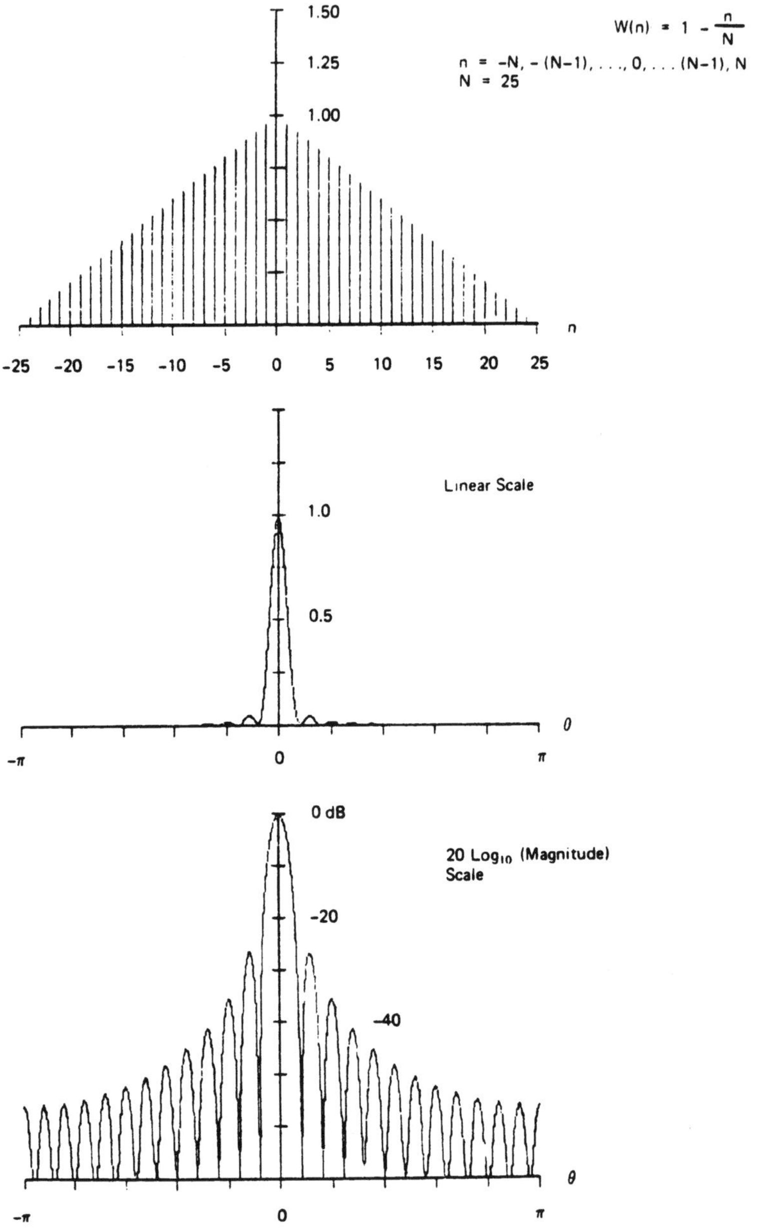

FIGURE 3A. Triangle (Bartlet) window. Upper trace, the window in the time domain; middle trace, the window in the frequency domain, linear scale; lower trace, the window in the frequency domain, logarithmic scale. (From Harris, F. J., Trigonometric Transforms, A Unique Introduction to the FFT, Tech. Publ. DSP-005 (8-81), Spectral Dynamic Division, Scientific Atlanta, San Diego, 1981. With permission.)

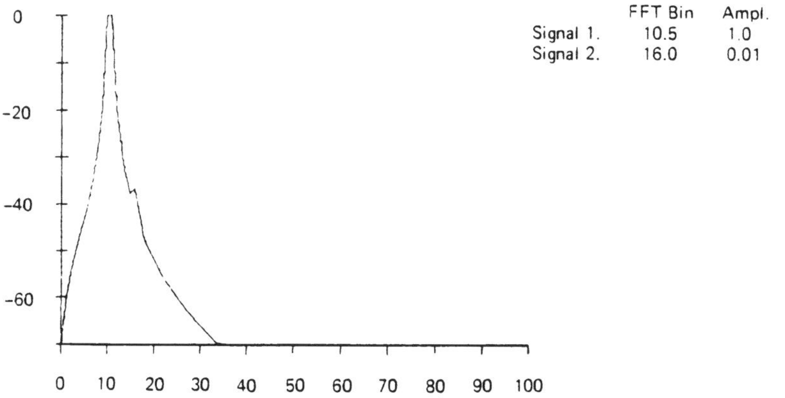

FIGURE 3B. Triangle (Bartlet) window. FFT power spectral density function estimation of synthesized signal consisting of two sinewaves with frequencies of 10.5 and 16 fs/N and amplitudes of 1.00 and 0.01, respectively. Data window was used (fs, sampling frequency; N, number of samples in the window). (From Harris, F. J., Trigonometric Transforms, A Unique Introduction to the FFT, Tech. Publ. DSP-005 (8-81), Spectral Dynamic Division, Scientific Atlanta, San Diego, 1981. With permission.)

$$W(k) = 0.54\ D(k) + 0.23\left[D\left(k - \frac{2\Pi}{N}\right) + D\left(k + \frac{2\Pi}{N}\right)\right] \tag{B.11}$$

where D(k) is given by Equation B.9.

Figures 5A and B depict the window. The SLL is more than 40 dB down. Note also that because of the deep attenuation due to the missing side lobe, the separation between the PSD of the two sine waves is better than in previous windows. We can allow more kernels in the window to achieve better side lobe cancellation. The *Blackman-Harris window* is given by:

$$w(n) = 0.35875 - 0.48829\ \mathrm{Cos}\left(\frac{2\Pi}{N}\ n\right) + 0.14128\ \mathrm{Cos}\left(\frac{2\Pi}{N}\ 2n\right) - 0.01168\ \mathrm{Cos}\left(\frac{2\Pi}{N}\ 3n\right) \tag{B.12}$$

This window yields SLL of more than 90 dB down.

F. Dolph-Chebyshev Window

The window can be designed to yield minimum main lobe width for a given side lobe level. The optimal solution yields the *Dolph-Chebyshev window* which is given in the frequency domain by a complicated equation.[9] The Dolph-Chebyshev window is really a family of windows with a parameter β being the log of the ratio of main lobe level to side lobe level. Figure 6 depicts the window (with β = 3.0) and the spectrum of the two sine waves achieved by means of this window. Note the level of the side lobes—60 db (β = 3.0), and the nice separation of peaks of the 2 sine waves.

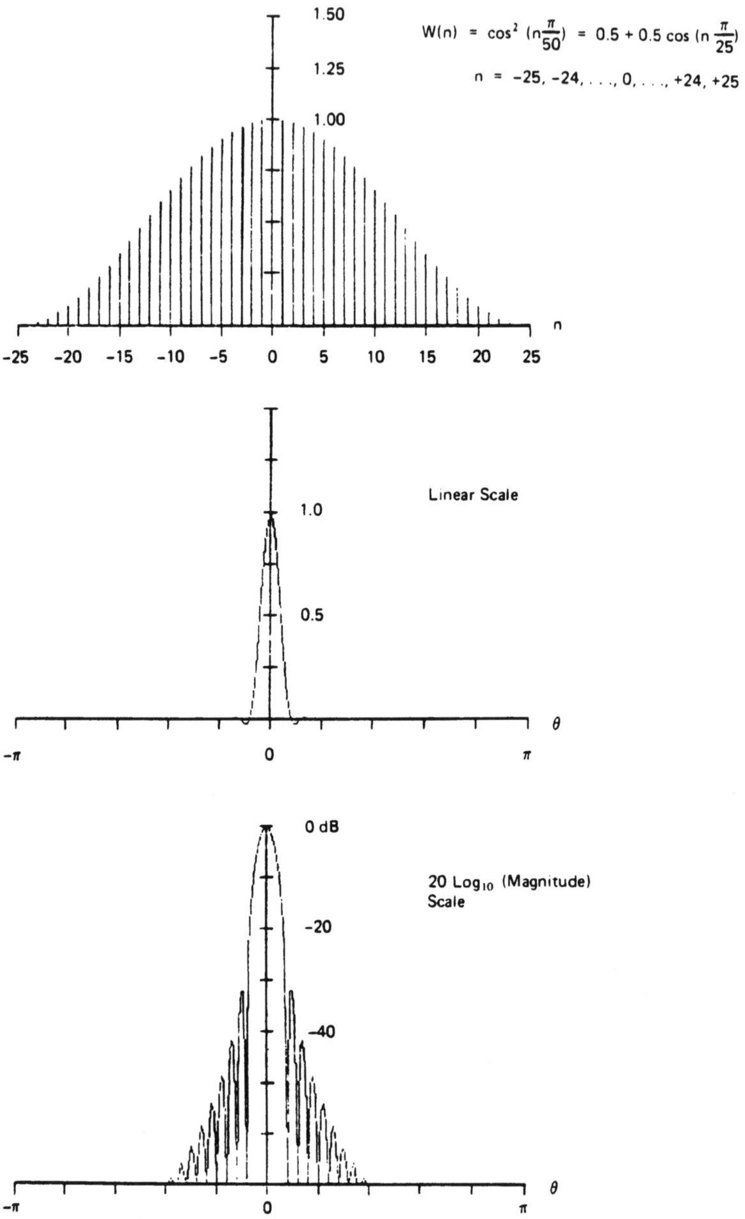

FIGURE 4A. The Hanning window, cosine$^{\alpha}$ window with $\alpha = 2$. Upper trace, the window in the time domain; middle trace, the window in the frequency domain, linear scale; lower trace, the window in the frequency domain, logarithmic scale. (From Harris, F. J., Trigonometric Transforms, A Unique Introduction to the FFT, Tech. Publ. DSP-005 (8-81), Spectral Dynamic Division, Scientific Atlanta, San Diego, 1981. With permission.)

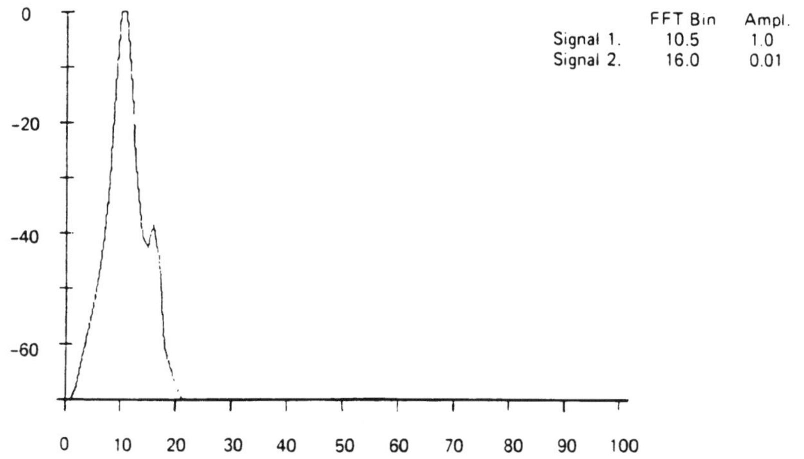

FIGURE 4B. The Hanning window, cosine$^{\alpha}$ window with $\alpha = 2$. FFT power spectral density function estimation of synthesized signal consisting of two sinewaves with frequencies of 10.5 and 16 fs/N and amplitudes of 1.00 and 0.01, respectively. Data window was used (fs, sampling frequency; N, number of samples in the window). (From Harris, F. J., Trigonometric Transforms, A Unique Introduction to the FFT, Tech. Publ. DSP-005 (8-81), Spectral Dynamic Division, Scientific Atlanta, San Diego, 1981. With permission.)

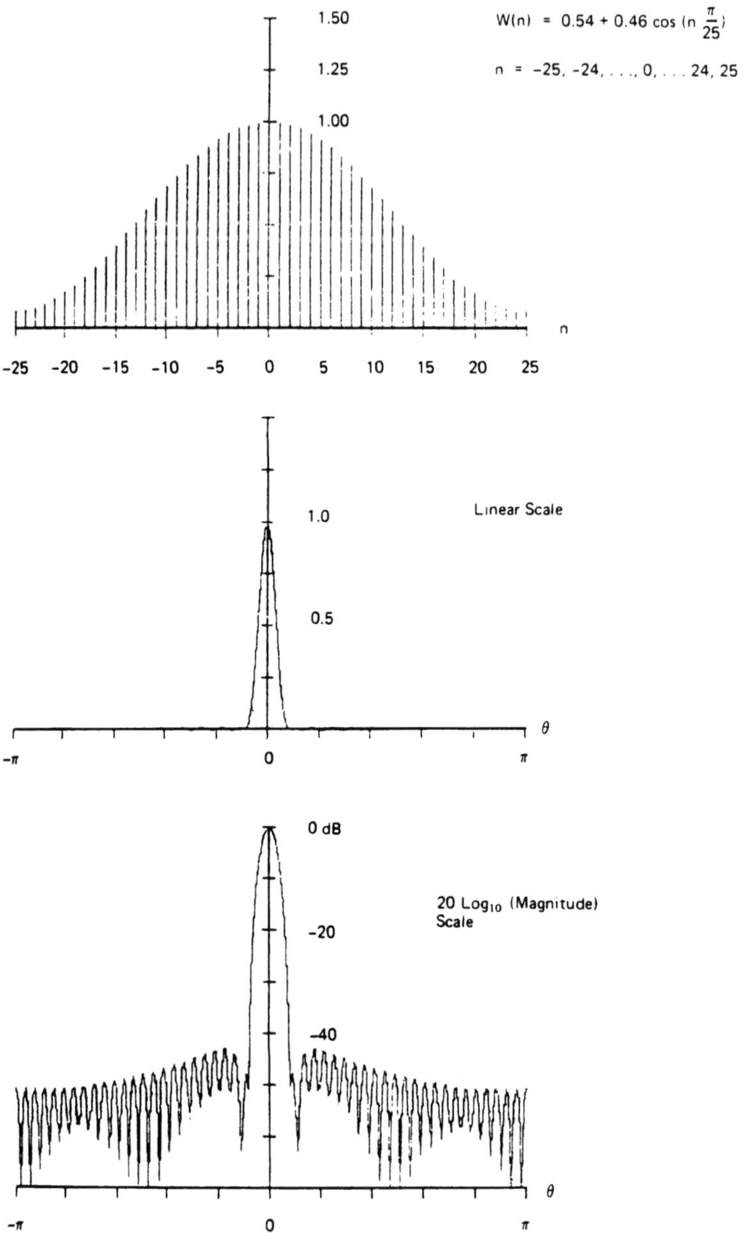

FIGURE 5A. The Hamming window. Upper trace, the window in the time domain; middle trace, the window in the frequency domain, linear scale; lower trace, the window in the frequency domain, logarithmic scale. (From Harris, F. J., Trigonometric Transforms, A Unique Introduction to the FFT, Tech. Publ. DSP-005 (8-81), Spectral Dynamic Division, Scientific Atlanta, San Diego, 1981. With permission.)

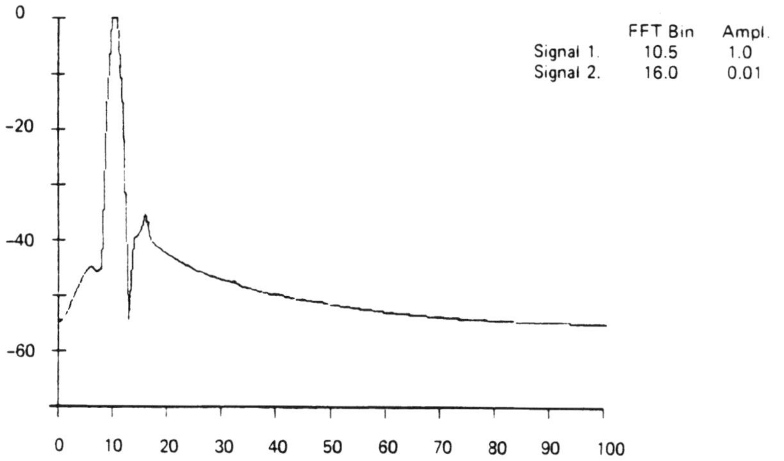

FIGURE 5B. The Hamming window. FFT power spectral density function estimation of synthesized signal consisting of two sinewaves with frequencies of 10.5 and 16 fs/N and amplitudes of 1.00 and 0.01, respectively. Data window was used (fs, sampling frequency; N, number of samples in the window). (From Harris, F. J., Trigonometric Transforms, A Unique Introduction to the FFT, Tech. Publ. DSP-005 (8-81), Spectral Dynamic Division, Scientific Atlanta, San Diego, 1981. With permission.)

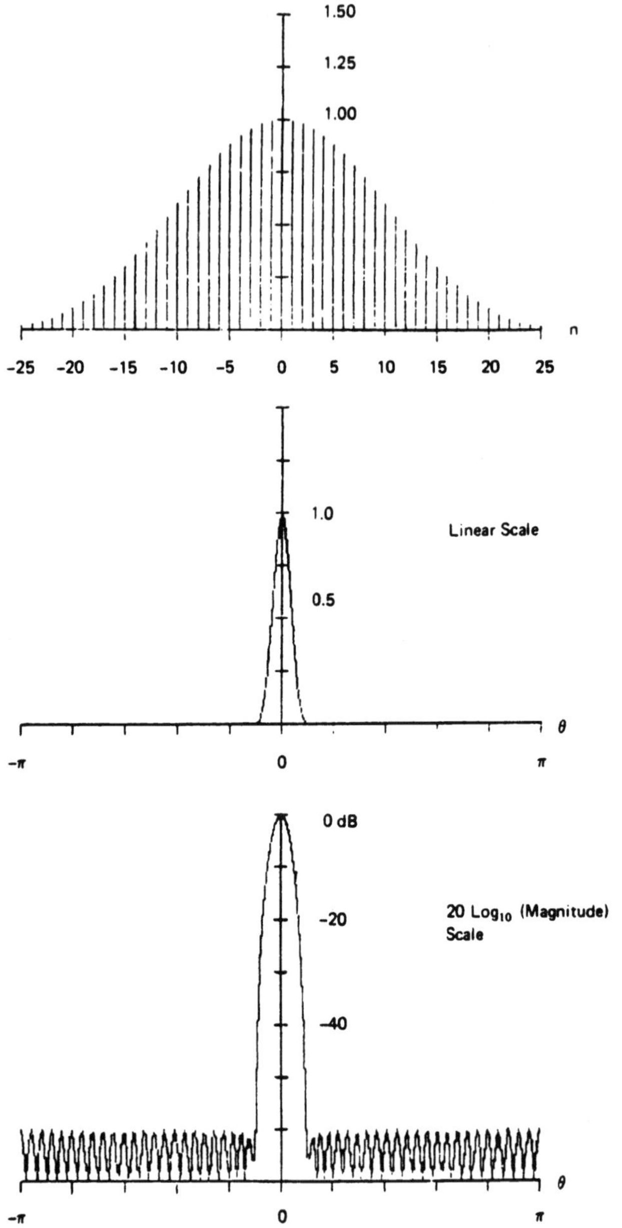

FIGURE 6A. The Dolph-Chebyshev window with $\beta = 3.0$. Upper trace, the window in the time domain; middle trace, the window in the frequency domain, linear scale; lower trace, the window in the frequency domain, logarithmic scale. (From Harris, F. J., Trigonometric Transforms, A Unique Introduction to the FFT, Tech. Publ. DSP-005 (8-81), Spectral Dynamic Division, Scientific Atlanta, San Diego, 1981. With permission.)

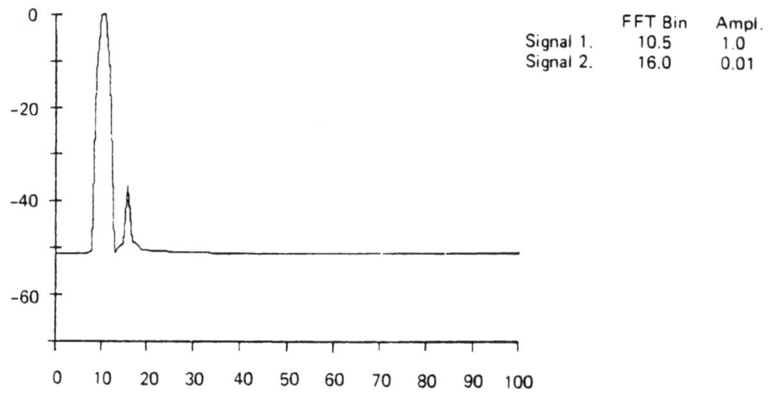

FIGURE 6B. The Dolph-Chebyshev window with $\beta = 3.0$. FFT power spectral density function estimation of synthesized signal consisting of two sinewaves with frequencies of 10.5 and 16 fs/N and amplitudes of 1.00 and 0.01, respectively. Data window was used (fs, sampling frequency; N, number of samples in the window). (From Harris, F. J., Trigonometric Transforms, A Unique Introduction to the FFT, Tech. Publ. DSP-005 (8-81), Spectral Dynamic Division, Scientific Atlanta, San Diego, 1981. With permission.)

REFERENCES

1. **Papoulis, A.,** *Signal Processing,* McGraw-Hill Int., Auckland, 1981.
2. **Gold, B. and Rader, C. M.,** *Digital Signal Processing,* McGraw-Hill, New York, 1969.
3. **Beauchamp, K. and Yuen, C.,** *Digital Methods for Signal Analysis,* George Allen and Unwin Ltd., London, 1979.
4. **Brillinger, D. R.,** *Time Series: Data Analysis and Theory,* Holden Day, San Francisco, 1981.
5. **Jenkins, G. M. and Watts, D. G.,** *Spectral Analysis and Its Applications,* Holden Day, San Francisco, 1968.
6. **Tretter, S. A.,** *Introduction to Discrete Time Signal Processing,* John Wiley & Sons, New York, 1976.
7. **Oppenheim, A. V. and Schafer, R. W.,** *Digital Signal Processing,* Prentice-Hall, Englewood Cliffs, N.J., 1975.
8. **Rabiner, L. R. and Gold, B.,** *Theory and Application of Digital Signal Processing,* Prentice-Hall, Englewood Cliffs, N.J., 1975.
9. **Harris, F. J.,** On the use of windows for harmonic analysis with DFT, *Proc. IEEE,* 66, 51, 1978.
10. **Nuttal, A. H.,** Some windows with very good sidelobe behavior, *IEEE Trans. Acoust. Speech Signal Process.,* 29, 84, 1981.
11. **Babic, H. and Temes, G. C.,** Optimum low-order windows for DFT systems, *IEEE Trans. Acoust. Speech Signal Process.,* 24, 512, 1976.
12. **Eberhard, A.,** An optimal discrete window for the calculation of power spectra, *IEEE Trans. Audio. Electroacoust.,* 21, 37, 1973.
13. **Rohling, H. and Schuermann, J.,** Discrete time window functions with arbitrary low sidelobe level, *Signal Process.,* 5, 127, 1983.
14. **Blomqvist, A.,** Figure of merit of windows for power spectral density spectrum estimation with the DFT, *Proc. IEEE,* 67, 438, 1979.
15. **Prabhu, K. M. M., Reddy, V. U., and Agrawal, J. P.,** Performance comparison of data windows, *Elect. Lett.,* 13, 600, 1977.
16. **Geckilini, N. C. and Yavuz, D.,** Some novel windows and concise tutorial comparison of window families, *IEEE Trans. Acoust. Speech Signal Process.,* 26, 501, 1978.
17. **Harris, F. J.,** Trigonometric Transforms, A Unique Introduction to the FFT, Tech. Publ. DSP-005 (8-81), Spectral Dynamic Division, Scientific Atlanta, San Diego, 1981.
18. **Brillinger, D. R.,** The key role of tapering in spectrum estimation, *IEEE Trans. Acoust. Speech Signal Process.,* 29, 1075, 1981.
19. **Yuen, C. K.,** Quadratic windowing in the segment averaging method for power spectrum computation, *Technometrics,* 20, 195, 1978.
20. **Yuen, C. K.,** On the smoothed periodogram method for spectrum estimation, *Signal Process.,* 1, 83, 1979.

Appendix C

COMPUTER PROGRAMS

I. INTRODUCTION

This appendix contains a number of computer programs and subroutines for biomedical signal processing. The programs are written in FORTRAN IV language and are used on the VAX 11/750 computer, under the VMS operating system. The input to the various programs are vectors containing the samples of the signals to be processed. These are read from data files generated by the A/D converter. The files are therefore unformatted, integer files. In order to make the software as compatible as possible with other machines, input and output statements, file definitions, and structuring are done with input-output subroutines (RFILE, WFILE, RFILEM, WFILEM). The user can adapt the software to another computer or use different data files by just replacing these subroutines.

The programs use several mathematical subroutines mainly for matrix operations. All of these subroutines are given in this appendix except for the subroutine EIGEN, which computes the eigenvalues and corresponding eigenvectors of a matrix. The listing of this subroutine was not included in the appendix due to its length. Subroutines for eigenvalues and eigenvectors computations can be found in one of the well known software libraries such as the IBM System/360 Scientific Subroutine Package (SSP), the International Mathematical and Statistical Libraries (IMSL), or the CERN Library.

The programs presented here are taken from the Bio-Medical Signal Processing Package (BMSPP) of the Center for Bio-Medical Engineering, Ben Gurion University. The listing of the complete package could not be presented here, due to space limitations. The few programs presented here were selected to allow the interested reader to implement some of the processing methods discussed in this book.

II. MAIN PROGRAMS

```
      PROGRAM NUSAMP
                       (VAX VMS VERSION)


      THIS PROGRAM PROVIDES THREE TYPES OF NON UNIFORM
      SAMPLING WITH APPLICATIONS TO BIOMEDICAL SIGNALS.
      DATA IS READ FROM UNFORMATTED INTEGER FILE.
      THE USER HAS A CHOICE OF ONE OUT OF THREE NON UNIFORM
      SAMPLING METHODS FOR DATA COMPRESSION:

            A. ZERO ORDER ADAPTIVE SAMPLING (VOLTAGE
                        TRIGGERING METHOD).

            B. FIRST ORDER ADAPTIVE SAMPLING (TWO POINTS
                        PROJECTION METHOD).

            C. SECOND ORDER ADAPTIVE SAMPLING (SECOND
                        DIFFERENCE METHOD).


      INPUT FILES:    UNFORMATTED,INTEGER FILE WITH NREC=NO.
                      OF RECORDS AND NOP=NO. OF SAMPLES PER RECORD.
                      EACH RECORD CONTAINS SAMPLES OF SIGNAL TO BE
                      COMPRESSED BY NON UNIFORM SAMPLING.

      OUTPUT FILES:   UNFORMATTED,INTEGER, FILE WITH 3 RECORDS AND
                      NOP SAMPLES IN EACH RECORD.
                          RECORD 1:THE ORIGINAL SIGNAL
                          RECORD 2:LOCATIONS OF NON UNIFORM SAMPLES
                                   (A VALUE OF 512 IS PLACED AT EACH
                                    SAMPLING LOCATION)
                          RECORD 3:THE RECONSTRUCTED SIGNAL

      REFERENCES:

            1. COHEN,A., BIOMEDICAL SIGNAL PROCESSING,
                     CRC PRESS, CHAPTER 4

            2. BLANCHARD,S.M. AND BARR,R.C.,ZERO FIRST AND
                     SECOND ORDER ADAPTIVE SAMPLING FROM ECGS
                     PROC. OF THE 35TH ACEMB,PHILADELPHIA,1982,209
C
C           3. PAHLM,O.,BORJESSON,P.D.AND WERNER,O.,
C                    COMPACT DIGITAL STORAGE OF ECGS, COMPUTER PROG.
C                    IN BIOMED.,9,293,1979
C
C
C
C
C      LINKING: RFILE
C
C


      INTEGER IVEC(2048),K(1024),KP(2048),IAVER(2),IREC(2048),
     *          IAUX(2048)
      REAL AVER(2)
      BYTE NAME(11),NAME1(11)



C
C               READ INPUT FILE
C
      CALL RFILE(NAME,IVEC,NOP,IAUX)
      TYPE 209
209   FORMAT(1H$'ENTER NO. OF POINTS FOR AVERAGING WINDOW: ')
      ACCEPT *,IAW
      IAW2=IAW*2
      TYPE 207
```

```
207     FORMAT(2X'THIS PROGRAM SAMPLES THE SIGNAL IN THE FILE'/2X
     *  'BY ONE OF THREE NON UNIFORM ADAPTIVE SAMPLING METHODS:'/10X
     *  '0-ZERO ORDER(VOLTAGE TRIGGER)-TYPICAL THRES.=0.025'/10X
     *  '1-FIRST ORDER (TWO POINTS PROJECTION)-THRES.=0.0003'/10X
     *  '2-SECOND ORDER (SECOND DIFFERENCE)-TYPICAL THRES.=0.0002'/)
        TYPE 208
208     FORMAT(1H$'ENTER TYPE OF SAMPLING (0,1,2): ')
        ACCEPT *,ISM
        TYPE 210
210     FORMAT(1H$'ENTER THRESHOLD LEVEL (IN VOLTS): ')
        ACCEPT *,R
        RMOD=R*1024
        IF (ISM-1) 10,40,70
C
C****** THE VECTOR K IS A VECTOR IN WHICH THE INDICES OF THE SAMPLES
C                ARE PLACED.****************
C
C
********************** ZERO ORDER (VOLTAGE TRIGGERED) METHOD*******
C
C       IF PREVIOUS SAMPLING POINT OCCURRED AT INDEX K(I) THE NEXT
C       SAMPLING,AT K(I+1), WILL BE DETERMINED BY:
C                ABS(X(K(I+1))-X(K(I))).GT.R
C        WHERE R IS THE THRESHOLD.
C

10      CONTINUE

C
C           OPEN AVERAGING WINDOW
C
        JJ=0
        REF=0
        DO 14 I=1,IAW
14      REF=REF+IVEC(I)
        REF=REF/IAW
        XFIRST=REF       !INITIAL CONDITION TO BE SENT FOR RECON.
        DO 11 I=1,(NOP-IAW),IAW
        IAVER(1)=0
        AVER(1)=0
        DO 12 II=1,IAW
12      AVER(1)=AVER(1)+IVEC(II+I-1)
        AVER(1)=AVER(1)/IAW
        IF (ABS(AVER(1)-REF).LE.RMOD) GO TO 11
C
C         SAMPLING POINT IS NEEDED
C
        JJ=JJ+1
        REF=AVER(1)
        K(JJ)=I+IAW
11      CONTINUE
C
C        PREPARING OUTPUT FOR PLOTING-RECONSTRUCTION OF SIGNAL
C
        JJ=1
        IREC(1)=XFIRST
        KP(1)=512
        DO 701 II=2,NOP
        KP(II)=0
        IF(K(JJ).NE.II) GO TO 700
        KP(II)=512
        IREC(II)=IVEC(II)
        JJ=JJ+1
        GO TO 701
700     IREC(II)=IREC(II-1)
701     CONTINUE
        GO TO 777
C
40      CONTINUE
C
```

```
C***************** FIRST ORDER (TWO POINTS PROJECTION )*******
C
C                       IF PREVIOUS SAMPLING POINT OCCURRED AT
C               INDEX K(I) THE NEXT SAMPLING ,AT K(I+1), WILL
C               BE DETERMINED BY:
C                          ABS(XDOT(K(I+1)-XDOT(K(I))).GT.R
C               WHERE XDOT IS THE ESTIMSTION OF THE DERIVATIVE
C               AND R IS A DERIVATIVE THRESHOLD.
C
C
        JJ=0
        AVER(1)=0
        AVER(2)=0
        DO 15 I=1,IAW
        AVER(1)=AVER(1)+IVEC(I)
 15     AVER(2)=AVER(2)+IVEC(I+IAW)
        AVER(1)=AVER(1)/IAW
        AVER(2)=AVER(2)/IAW
        REF=(AVER(2)-AVER(1))/(IAW-1)
C
C        REF IS THE ESTIMATE OF THE INITIAL (SMOOTHED) DERIVATIVE
C
        AVER(1)=AVER(2)
        DO 16 I=IAW2,(NOP-IAW),IAW
        AVER(2)=0
        DO 17 II=1,IAW
 17     AVER(2)=AVER(2)+IVEC(II+I)
        AVER(2)=AVER(2)/IAW
        XDOT=(AVER(2)-AVER(1))/(IAW-1)
        IF(ABS(REF-XDOT).LE.RMOD) GO TO 801
C
C         SAMPLING POINT IS NEEDED
C
        JJ=JJ+1
        REF=XDOT
        K(JJ)=I+IAW
        KP(I+IAW)=512
801     DO 800 II=I,(IAW+I)
800     IREC(II)=IREC(II-1)+REF
16      AVER(1)=AVER(2)
        GO TO 777
70      CONTINUE
C
*********** SECOND ORDER-SECOND DIFFERENCE METHOD
C
C                        IF PREVIOUS SAMPLING POINT OCCURRED AT
C            INDEX K(I) THE NEXT SAMPLING POINT ,AT INDEX K(I+1),
C            WILL BE DETERMIND BY:
C
C                        ABS(XDOT(K(I+1))-XDOT(K(I+1)-1)).GT.R
C
C            WHERE XDOT IS THE ESTIMATE OF THE DERIVATIVE AND
C            R IS THE THRESHOLD.
C
        JJ=0
        AVER(1)=0
        AVER(2)=0
        DO 45 I=1,IAW
        AVER(1)=AVER(1)+IVEC(I)
 45     AVER(2)=AVER(2)+IVEC(I+IAW)
        AVER(1)=AVER(1)/IAW
        AVER(2)=AVER(2)/IAW
        XDOTP=(AVER(2)-AVER(1))/(IAW-1)
        AVER(1)=AVER(2)
        DO 46 I=IAW2,(NOP-IAW),IAW
        AVER(2)=0
        DO 47 II=1,IAW
 47     AVER(2)=AVER(2)+IVEC(II+I)
        AVER(2)=AVER(2)/IAW
        XDOT=(AVER(2)-AVER(1))/(IAW-1)
        IF (ABS(XDOT-XDOTP).LE.RMOD) GO TO 601
```

```
C
C          SAMPLING POINT IS NEEDED
C
        JJ=1+JJ
        K(JJ)=I+IAW
        KP(I+IAW)=512
        XDOTP=XDOT
601     DO 600 II=I,(IAW+I)
600     IREC(II)=IREC(II-1)+XDOTP
        AVER(1)=AVER(2)
46      XDOTP=XDOT
C
C
C************ OUTPUT FILE HAS 3 RECORDS:
C                       1. ORIGINAL SIGNAL
C                       2. LOCATIONS OF NON UNI. SAMPLES.
C                       3. RECONSTRUCTED SIGNAL
C
C
777     CONTINUE
C
C
C               WRITE RESULTS ON OUTPUT FILE
C
        TYPE 211
211     FORMAT(1H$'ENTER OUTPUT FILE NAME: ')
        ACCEPT 119,NCHO,(NAME1(I),I=1,11)
119     FORMAT(Q,11A1)
        NOP2=NOP*2
        CALL ASSIGN (2,NAME1,11)
        DEFINE FILE 2(3,NOP2,U,IVAR)
        WRITE(2'1)(IVEC(II),II=1,NOP)
        WRITE(2'2)(KP(II),II=1,NOP)
        WRITE(2'3)(IREC(II),II=1,NOP)
        CALL CLOSE(2)
C
C               PRINT PROGRAM'S STATISTICS
C
        PRINT 900
900     FORMAT(/25X'RESULTS OF NUSAMP PROGRAM')
        PRINT 901
901     FORMAT(25X'***************************'/)
        PRINT 902,(NAME(I),I=1,11)
902     FORMAT(15X'INPUT FILE NAME: ',11A1)
        PRINT 905,IAW
905     FORMAT(25X'NO. OF SAMPLES IN AVERAGING WINDOW= 'I4)
        PRINT 908,R
908     FORMAT(25X'THRESHOLD LEVEL= 'E10.3)
        PRINT 901
        PRINT 906,ISM
906     FORMAT(15X'NON UNIFORM SAMPLING OF ORDER 'I1)
        PRINT 907,(NAME1(I),I=1,11)
907     FORMAT(/15X'NAME OF OUTPUT FILE: '11A1)
        PRINT 909,(JJ-1)
909     FORMAT(25X'NO. OF SAMPLES USED= 'I6)
C
C            COMPRESSION RATIO (CR) IS THE RATIO BETWEEN
C            NO. OF SAMPLES (12 BITS) OF ORIGINAL SIGNAL
C            AND NO. OF SAMPLES OF NON UNIFORMALY SAMPLED



C            SIGNAL (JJ) PLUS THE LOCATIONS OF THE SAMPLES
C            (THESE ARE QUANTIZED AT 8 BITS)
C
        IF (JJ.NE.0) CR=NOP*12/(JJ*20)
        PRINT 910,CR
910     FORMAT(25X'COMPRESSION RATIO = 'E10.3)
        PRINT 901
        STOP
        END
```

```
      PROGRAM SEGMNT
C                    (VAX VMS VERSION)
C
C
C               THIS PROGRAM PROVIDES ADPTIVE SEGMENTATION OF
C               A SAMPLED FUNCTION. SEGMENTATION IS PERFORMED BY
C               ESTIMATING AN AR FILTER FOR AN INITIAL REFERENCE
C               WINDOW OF THE SIGNAL. THE INVERSE OF THIS FILTER
C               (THE WHITENING FILTER) IS USED TO WHITEN THE SAMPLES
C               OF A SLIDING WINDOW CONTINUOSLY RUNNING ALONG THE
C               TIME AXIS. THE WHITNNES OF THE RESIDUALS (THE OUTPUT
C               OF THE FILTER) IS EXAMIND. AS LONG AS THE RESIDUALS
C               ARE CONSIDERED WHITE THE CORRESPONDING SIGNAL WINDOW
C               BELONGS TO THE PREVIOUS SEGMENT. WHEN THE WHITNNES
C               MEASURE (SEM) CROSSES A GIVEN THRESHOLD- THE
C               WINDOW IS CONSIDERED BELONGING TO A NEW SEGMENT.
C               A NEW REFERENCE WINDOW IS DEFINED AND THE PROCESS
C               CONTINUES.
C
C
C       INPUT:
C
C               1.  UNFORMATTED INTEGER DATA FILE TO BE SEGMENTED
C                   WITH NOR RECORDS AND NOSR SAMPLES PER RECORD.
C
C
C
C       OUTPUTS:
C
C               1.  UNFORMATTED INTEGER FILE HOLDING THE SEM FUNCTION
C                   WITH NOR RECORDS AND NOSR SAMPLES PER RECORD.
C                   (NAME XXXXX GIVEN BY USER)
C
C               2.  FORMATTED INTEGER FILE HOLDING THE AR ORDER,
C                   THE NO. OF SEGMENTS (ISEG) AND THE INDICES
C                   OF THE SEGMENTS-FILE NAME:
C                   ( THREE FIRST CHARACTERS OF IN FILE).IND
C
C               3. FORMATTED REAL FILE HOLDING THE LPC AND PAR
C                  OF EACH SEGMENT-FILE NAME:
C                  ( THREE FIRST CHARACTERS OF IN FILE).LPC
C
C
C       LINK: NACOR,DLPC,RFILE,WFILE
C
C
C       REFERENCE:
C
C               1. COHEN,A. BIOMEDICAL SIGNAL PROCESSING,
C                        CRC PRESS, CHAPTER 7
C
C               2. BODENSTEIN,G. AND PRAETORIUS,H.M. FEATURE EXTRACTION
C                        FROM EEG BY ADAPTIVE SEGMENTATION, PROC. IEEE.
C                        65,642,1977
C
C
C
      DIMENSION COR(41),AUX(41),SMPR(1024),CORRES(41)
     *            ,RES(1024)
      REAL LPC(41),PAR(41)
      BYTE NAME(11),NAME1(11),NAME2(11)
      INTEGER ISMP(12288),INDC(1000),ISEM(12288),IAUX(2048)

C
C               READ INPUT FILE
C
      CALL RFILE(NAME,ISMP,NTS,IAUX)
100   CONTINUE
C
C
C               OPEN LPC &PAR OUTPUT FILE
```

```
        NAME(4)='.'
        NAME(5)='L'
        NAME(6)='P'
        NAME(7)='C'
        CALL ASSIGN (2,NAME,7)
731     FORMAT(2E12.5)
C
C
C               CALCULATE LPC OF REFERENCE WINDOW
C               REFERENCE WINDOW HAS 256 SAMPLES
C               AND A FILTER OF ORDER INN (MAX. 40) IS USED
C
C
C
        ISEG=0      ! ISEG IS THE NO. OF CURRENT SEGMENT
        ICW=1       ! ICW IS THE INDEX OF CURRENT SAMPLE
        IBR=1       ! IBR IS THE INDEX OF CURRENT REFERENCE WINDOW
        INDC(1)=1   ! INDC IS THE VECTOR IN WHICH THE SEGMENTS
C                     INDICES ARE STORED (=IBR)
        IW=128      ! IW IS THE NO. OF SAMPLES IN THE REFERENCE
C                     AND SLIDING WINDOWS
        INN=20      ! INN IS THE ORDER OF THE WHITENING FILTER
C                     (MAX. 40)
        SEMTH=5. ! SEMTH IS THE THRESHOLD FOR SEM
        RES(1)=0.0
        IWNN=IW+INN
        INN1=INN+1
C
C               BEGIN CALCULATIONS OF THE ISEG'TH  REFERENCE WINDOW
C
103     CONTINUE
        ISEG=ISEG+1
        TYPE 555,ISEG,ICW
555     FORMAT(X'PROCESSING SEGMENT NO.: 'I3' SAMPLE NO.: 'I5)
        IBR=ICW
        INDC(ISEG)=IBR
        DO 102 I=1,IW
102     SMPR(I)=ISMP(IBR+I-1)
        CALL NACOR(SMPR,IW,COR,INN1,ENG)
        CALL DLPC(INN,COR,LPC,PAR,AUX,ERR)
C
C               LPC HOLDS THE WHITENING FILTER OF THE ISEG'TH
C               REFERENCE WINDOW
C
C               WRITE LPC AND PAR ON OUTPUT FILE
C
        DO 730 I=1,INN
730     WRITE(2,731) (LPC(I),PAR(I))
C
C               PREPARE DATA VECTOR FOR FIRST SLIDING WINDOW
C
        ISW=IBR+IW    ! ISW IS THE INDEX OF THE FIRST SLIDING WINDOW
        DO 701 I=1,IW
701     SMPR(I)=ISMP(ISW+I-1)
        ICW=ISW-1
C
C               ZERO SEM VECTOR IN REFERENCE WINDOW
C
        DO 717 I=IBR,ICW
717     ISEM(I)=0
104     CONTINUE
C
C               START CHECKING SLIDING WINDOWS IN CURRENT SEGMENT
C               UPDATE SLIDING WINDOW
        ICW=ICW+1      !ICW IS THE INDEX OF THE END OF CURRENT
C                       SLIDING WINDOW (CURRENT SAMPLE)
        IF(ICW.GE.NTS) GO TO 711   ! END OF DATA VECTOR
        DO 702 I=1,IW-1
702     SMPR(I)=SMPR(I+1)
        SMPR(IW)=ISMP(ICW+IW)
        IF(ICW.GT.(ISW+IW)) GO TO 705
```

```
C
C               CALCULATE THE RESIDUALS OF FIRST SLIDING WINDOW
C
        DO 720 I=1,INN-1
        RES(I)=SMPR(I)
        DO 720 J=1,I-1
720     RES(I)=RES(I)+SMPR(I-J)*LPC(J)
        DO 700 I=INN,IW
        RES(I)=SMPR(I)
        DO 700 J=1,INN
        RES(I)=RES(I)+SMPR(I-J)*LPC(J)
700     CONTINUE
C
C               CALCULATE CORRELTIONS OF RES
C
C               FIND CORRELATIONS FOR FIRST SLIDING WINDOW
C
        CALL NACOR(RES,IW,CORRES,INN,ENGRES)

        DO 708 I=1,INN1
708     CORRES(I)=CORRES(I)*ENGRES
        CTH=0.9*ENGRES
        GO TO 706
C
C               CALCULATE NEW RESIDUAL
C
705     CONTINUE
        RESW=SMPR(IW)
        DO 733 J=1,INN
733     RESW=RESW+SMPR(IW-J)*LPC(J)
        CORRES(1)=CORRES(1)-RES(1)*RES(1)+RESW*RESW
C
C               FIND CORRELATIONS ITERATIVELY FOR ALL SLIDING
C               WINDOWS EXCEPT THE FIRST ONE
        DO 707 J=2,INN1
707     CORRES(J)=CORRES(J)-RES(1)*RES(J)+RES(IW-J+2)*RESW
C
C               SHIFT RESIDUALS VECTOR
C
        DO 734 I=1,IW-1
734     RES(I)=RES(I+1)
        RES(IW)=RESW
706     CONTINUE
C
C               CLIP CORRELATIONS TO REMOVE SHORT TRANSIENT
C               ARTIFACTS
C
        DO 709 I=1,INN1
709     IF(CORRES(I).GT.CTH) CORRES(I)=CTH
C
C               CALCULATIONS OF SEM
C
        SUM=0.0
        DO 710 I=2,INN1
710     SUM=SUM+(CORRES(I))*(CORRES(I))
        SEM=(ERR/CORRES(1)-1)**2+2*SUM/(CORRES(1)*CORRES(1))
        ISEM(ICW)=INT((SEM*409.6)+0.5)
C
C               COMPARE SEM WITH THRESHOLD
C
        IF (SEM.GT.SEMTH) GO TO 103       ! START A NEW SEGMENT
        GO TO 104                         ! STAY IN CURRENT SEGMENT
C                                           SHIFT SLIDING WINDOW
711     CONTINUE                          ! END OF DATA VECTOR
C
C               CLOSE LPC OUTPUT FILE
C
        CALL CLOSE (2)
C
C               OPEN AND WRITE OUTPUT SEM FILE
```

```
C
        CALL WFILE(NAME1,ISEM,NTS,NORS)
        IVER=2
C       PRINT 722,IVER
C722    FORMAT(20X,'RESULTS OF SEGMNT PROGRAM....VERSION: 'I2)
        PRINT 723,(NAME(I),I=1,11)
723     FORMAT(5X'INPUT FILE NAME: '11A1)
        PRINT 725,ISEG
725     FORMAT(/5X'DATA WAS DIVIDED INTO 'I3' SEGMENTS')
        PRINT 726,(NAME1(I),I=1,11)
726     FORMAT(/5X'THE SEM VECTOR IS STORED IN FILE: '11A1)
        NAME(5)='I'
        NAME(6)='N'
        NAME(7)='D'
        CALL ASSIGN (2,NAME,7)
        WRITE(2,727),INN,ISEG
727     FORMAT(2I5)
        DO 728 I=1,ISEG
728     WRITE(2,727) (INDC(I))
        CALL CLOSE (2)
999     CONTINUE
        STOP
        END
```

```
C
        NLS=NTS+NZPAD
        N=1
        NP=0
1       N=N*2
        NP=NP+1
        IF(N.LT.NLS) GO TO 1
        TYPE 105,NLS
105     FORMAT(X'LENGTH OF PADDED DATA VECTOR(POWER OF 2): 'I4)
        IF(NLS.GT.2048) TYPE 109
109     FORMAT(X'MAX. LENGTH OF DATA VECTOR IS 2048 !!')
C
C
        DO 3 II=1,NTS
3       COR(II)=ISAMP(II)
C
C
C       ZERO PADDING
C
        DO 5 I=NTS+1,NLS
5       COR(I)=0.0
C
C       FFT CALCULATIONS
C
        CALL FT01A(NLS,2,COR,CORI)
        NLSH=NLS/2
        DO 6 I=1,NLSH
6       COR(I)=SQRT(COR(I)*COR(I)+CORI(I)*CORI(I))
C
C********* NORMALIZATION OF ESTIMATED PSD FUNCTION *********
C
        CALL XTERM(COR,NLSH,CMAX,U)
        DO 7 I=1,NLSH
7       ISAMP(I)=INT((COR(I)/CMAX)*1024+0.5)
C
C********* OUTPUT PROCEDURES *********
C
        CALL WFILE(NAME1,ISAMP,NLSH,NORO)
        PRINT 110
110     FORMAT(20X'RESULTS OF PROGRAM PERSPT-'/25X
     *  'ESTIMATION OF PSD BY THE PERIODOGRAM')
        PRINT 111,(NAME(I),I=1,11)
```

```
111     FORMAT(/10X'INPUT FILE NAME: '11A1)
        PRINT 112,NTS
112     FORMAT(10X'TOTAL NO. OF SAMPLES: 'I4)
        PRINT 113,NZPAD
113     FORMAT(10X'NO.OF PADDING ZEROES: 'I4)
        PRINT 114,(NAME1(I),I=1,11),NLSH
114     FORMAT(/10X'OUTPUT FILE NAME: '11A1/
     *  10X'NO. OF RECORDS: 1  NO. OF SAMPLES: 'I4)
        STOP
        END




C
        PROGRAM PERSPT
C                         (VAX VMS VERSION)
C
C
C
C                    THIS PROGRAM ESTIMATES THE PSD FUNCTION
C                    BY MEANS OF THE PERIODOGRAM (ABS. VALUE OF
C                    WINDOWED DATA RECORD FFT).THE DATA RECORD
C                    IS AUGMENTED BY PADDING ZEROES AS REQUESTED
C                    BY THE USER.
C
C
C       REFERENCE:
C                  1.  COHEN,A., BIOMEDICAL SIGNAL PROCESSING
C                            CRC PRESS, CHAPTER 8
C                  2.  OTENS,R.K. AND ENOCHSON,L., DIGITAL
C                            TIME SERIES ANALYSIS WILEY,1972
C
C
C
C       INPUT:
C                    UNFORMATTED INTEGER DATA FILE
C                    NO. OF RECORDS AND SAMPLES DETERMIND
C                    BY USER
C
C       OUTPUT:
C                    UNFORMATTED,INTEGER FILE WITH ONE RECORD
C                    STORING THE NORMALISED PSD ESTIMATIONS
C
C
C       LINKING:
C                    FT01A,XTERM,RFILE,WFILE
C
C
        INTEGER ISAMP(2048),IAUX(2048)
        REAL SAMP(4096),COR(2048),CORI(2048)
        BYTE NAME(11),NAME1(11)
C
C               READ INPUT FILE
C
        CALL RFILE(NAME,ISAMP,NTS,IAUX)
        TYPE 102
102     FORMAT(1H$'GIVE NO. OF PADDING ZEROES: ')
        ACCEPT *,NZPAD
C
C       CHECK FOR POWER OF TWO

        PROGRAM WOSA
C                    (VAX VMS VERSION)
C
C
C
C       THIS PROGRAM ESTIMATES THE POWER SPECTRAL
C       DENSITY FUNCTION (PSD) OF A STOCHASTIC PROCESS
C       BY MEANS OF THE 'WOSA' METHOED. THE TIME SAMPLE
```

```
C        FUNCTION PROCESS IS DIVIDED INTO SEGMENTS (IN GENERAL
C        OVERLAPPING SEGMENTS) EACH SEGMENT IS WINDOWED BY
C        MEANS OF RECTANGULAR,TRIANGULAR  OR HAMMING WINDOW ITS PSD
C        IS ESTIMATED BY THE SQUARE ABSOLUTE OF THE FFT.
C        THE FINAL PSD IS ESTIMATED BY AVERAGING ALL THE
C        SEGMENTAL PSDS.
C                THIS METHOED IS KNOWN AS WINDOWED OVERLAPPING
C        SPECTRUM ANALISIS (WOSA).
C
C        REFERENCES:
C
C                1. COHEN,A. BIOMEDICAL SIGNAL PROCESSING,
C                   CRC PRESS, CHAPTER 8
C
C                2. WELCH,P.D. THE USE OF FAST FOURIER TRANSFORM
C                   FOR THE ESTIMATION OF POWER SPECTRA: A METHOD
C                   BASED ON TIME AVERAGING OVER SHORT MODIFIED
C                   PERIODOGRAMS, IEEE TRAS. AUDIO ELECTRO.
C                   ,AU-15,70,1967
C
C
C                THE REQUESTED INFORMATION IS:
C                        (1) INPUT FILE NAME.
C                        (2) NUMBER OF SAMPLES IN THE INPUT FILE.
C                        (3) LENGTH OF SEGMENT (POWER OF 2) FOR FFT.
C                        (4) PERCENTAGE OF OVERLAP BETWEEN SEGMENTS.
C                        (5) TYPE OF WINDOW : (a) rectangular window
C                                             (b) triangular window
C                                             (c) hamming window
C                        (6) OUTPUT FILE NAME.
C
C        INPUT FILE:
C
C                UNFORMATTED INTEGER DATA FILE WITH NOR RECORDS
C                AND NOSR SAMPLES
C
C        OUTPUT FILE:

C                UNFORMATTED INTEGER FILE WITH THE AVERAGED
C                ESTIMATED PSD (ONE RECORD AND NLSH SAMPLES)
C
C
C        LINKING : FT01A,XTERM,RFILE,WFILE
C
C

         DIMENSION ISMP(16384),IAUX(2048)
         INTEGER PEROV,ISPACE,NEWSIZ,NOOREC,NOS,NLS,NLSP,NOREC
         BYTE NAME(13),AA(9),NAMEO(13)
         REAL FRE(2048),FIM(2048),SPCT(2048)
C
C
C
         DO 199 I=1,16384
         ISMP(I)=0
199      CONTINUE
C
C                READ INPUT FILE
C
C
         CALL RFILE(NAME,ISMP,NOS,IAUX)
         TYPE 3
3        FORMAT (1H$,'GIVE LENGTH OF SEGMENT (POWER OF 2) FOR FFT: ')
         ACCEPT*,NLS
C
C
C
C
C        CHECK IF POWER OF 2
C
         N=1
         NP=0
```

```
40      N=N*2
        NP=NP+1
        IF(N.LE.NLS) GOTO 40
        NLS=N/2
C
        TYPE 4,NLS
4       FORMAT (1X,'LENGTH OF SEGMENT (CLOSEST POWER OF 2) IS: ',I4)
        TYPE 5
5       FORMAT(1H$,'GIVE PERCENT OVERLAPPING BETWEEN SEGMENTS: ')
        ACCEPT*,PEROV
C
        TYPE 8
8       FORMAT(1H$,'GIVE WINDOW: RECTAN.=1; TRIA.=2; HAMMING=3; : ')
        ACCEPT*,IW
C
C
C       CALCULATING WITH PERCENTAGE OF OVERLAP (PEROV) AND SEGMENT
C       LENGTH NLS THE NO. OF SEGMENT AVAILABLE.
C
        NLSP=(NLS*PEROV)/100
        ITEMP=NLS
        NOREC=1
29      IF(ITEMP.GT.NOS) GOTO 30

        ITEMP=ITEMP-NLSP+NLS
        NOREC=NOREC+1
        GOTO 29
30      NOREC=NOREC-1       ! NOREC IS THE NO. OF OVERLAPPED SEGMENTS
C
C       CALCULATE FFT OF EACH SEGMENT AND AVERAGE
C
        NLSH=NLS/2
        IB=NLSP-NLS+2
        DO 197 I=1,NOREC
        IB=IB+NLS-1-NLSP
        DO 196 J=1,NLS
        FIM(J)=0.
196     FRE(J)=ISMP(IB+J-1)
        CALL FT01A(NLS,2,FRE,FIM)
        DO 195 J=1,NLSH
195     SPCT(J)=SPCT(J)+SQRT(FRE(J)*FRE(J)+FIM(J)*FIM(J))
197     CONTINUE
        DO 194 J=1,NLSH
194     SPCT(J)=SPCT(J)/NOREC
        CALL XTERM(SPCT,NLSH,SMAX,SMIN)
        DO 192 I=1,NLSH
192     ISMP(I)=INT((SPCT(I)/SMAX)*1024+0.5)
C
C       WRITE OUTPUT FILE
C
        CALL WFILE(NAMEO,ISMP,NLSH,NOR)
191     FORMAT(E12.5)
        CALL DATE(AA)
        PRINT 602,(AA(I),I=1,9)
602     FORMAT(/20X,'RESULTS OF 'WOSA'  PROGRAM , DATE:  ',9A1)
C
        PRINT 702
702     FORMAT(19X,'**************************************')
C
C
C
        PRINT 604,(NAME(I),I=1,11)
604     FORMAT(/10X,'*****INPUT ORIGINAL DATA FILE:  ',11A1)
C
C
        PRINT 607,PEROV
607     FORMAT(/2X,'PERCENTAGE OF OVERLAPP: ',I3)
C
        PRINT 608,NLSP
608     FORMAT(/2X,'NUMBER OF POINTS OVERLAPPED BETWEEN SEGMENTS: ',I4)
C
```

```
      PRINT 609,NOREC
609   FORMAT(/2X,'NUMBER OF SEGMENTS WILL BE USED IN 'WOSA': ',I4)
C
C***********************************************************************
C
      IF (IW-2) 610,620,630
C
610   PRINT 611
611   FORMAT(/5X,'TYPE OF WINDOW: RECTANGULAR')
      GO TO 640
C
620   PRINT 621
621   FORMAT(/5X,'TYPE OF WINDOW: TRIANGULAR')
      GO TO 640
C
630   PRINT 631
631   FORMAT(/5X,'TYPE OF WINDOW: HAMMING')
C
640   PRINT 641,(NAMEO(I),I=1,11)
641   FORMAT(/10X,'OUTPUT FILE NAME: ',11A1)
C
C
C
999   STOP
      END




      PROGRAM MEMSPT
C                    (VAX VMS VERSION)
C
C
C
C                 THIS PROGRAM ESTIMATES THE PSD FUNCTION
C                 BY MEANS OF THE MAXIMUM ENTROPY METHOD
C                 (MEM) OR THE AUTO REGRESIVE (AR) METHOD.
C
C
C     REFERENCE:
C                 1.  COHEN,A., BIOMEDICAL SIGNAL PROCESSING
C                          CRC PRESS, CHAPTER 8
C
C                 2.  BURG,J.P., MAXIMUM ENTROPY SPECTRAL ANALYSIS
C                     PROC. 37TH ANN. INST. MEETING SOC. EXPLOR.
C                     GEOPH. ,OKLAHOMA,1967
C
C
C     INPUT:
C                 UNFORMATTED INTEGER DATA FILE
C                 NO. OF RECORDS AND SAMPLES DETERMIND
C                 BY USER
C
C     OUTPUT:
C                 UNFORMATTED,INTEGER FILE WITH ONE RECORD
C                 STORING THE NORMALISED PSD ESTIMATIONS
C
C
C     LINKING:
C                 NACOR,DLPC,XTERM,RFILE,WFILE
C
C
      INTEGER ISAMP(2048),IAUX(2048)
      REAL SAMP(4096),COR(41),LPC(41),PAR(41),RHO(41),AUX(41)
      BYTE NAME(11),NAME1(11)
C
C          READ INPUT FILE
C
      CALL RFILE(NAME,ISAMP,NTS,IAUX)
```

```
      DO 3 I=1,NTS
3     SAMP(I)=ISAMP(I)
887   TYPE 103
103   FORMAT(H$'GIVE ORDER OF AR MODEL: ')
      ACCEPT *,NAR
      IF(NAR.LE.40) GO TO 888
      TYPE 109

109   FORMAT(X'MAX. ORDER OF AR MODEL IS 40 !!')
      GO TO 887
888   TYPE 886
886   FORMAT(1H$'GIVE ORDER OF SPECTRUM VECTOR: ')
      ACCEPT *,IT
C
C     CALCULATIONS OF CORRELATION
C
      NCORP1=NAR+1
      CALL NACOR(SAMP,NTS,COR,NCORP1,ENG)
C
C     ESTIMATIONS OF AR COEFFICIENTS
C
      CALL DLPC(NAR,COR,LPC,PAR,AUX,ERR)
C
C     PSD ESTIMATION
C                          CALCULATE RHO0 WITH A0=1
C
      RHO0=1.0
      DO 320 K=1,NAR
320   RHO0=RHO0+LPC(K)*LPC(K)
C
C                          CALCULATE RHOI
C
      DO 360 I=1,NAR-1
      RHOI=LPC(I)
      DO 340 J=1,NAR-I
340   RHOI=RHOI+(LPC(J))*(LPC(J+I))
      RHO(I)=RHOI
360   CONTINUE
      RHO(NAR)=LPC(NAR)
C
C     CALCULATE THE DISCRETE SPECTRUM
C
      PI2=8.0*ATAN(1.0)
      IT2=2*IT
      DO 400 K=1,IT
      SIGMA=0.
      DO 380 I=1,NAR
      SIGMA=SIGMA+(RHO(I))*COS(PI2*I*K/IT2)
380   CONTINUE
      SAMP(K)=ERR/(RHO0+2*SIGMA)
400   CONTINUE
C
C********* NORMALIZATION OF ESTIMATED PSD FUNCTION **********
C
      CALL XTERM(SAMP,IT,CMAX,CMIN)
      ACMIN=ABS(CMIN)
      IF(CMAX.LT.ACMIN) CMAX=ACMIN
      DO 7 I=1,IT
7     ISAMP(I)=INT((SAMP(I)/CMAX)*1024+0.5)
C
C********* OUTPUT PROCEDURES *********
C
      CALL WFILE(NAME1,ISAMP,IT,NORX)
      PRINT 110
110   FORMAT(20X'RESULTS OF PROGRAM MEMSPT-'/25X
```

```
      PROGRAM NOICAN
C                    (VAX VMS VERSION)
C
C               THIS PROGRAM USES THE LMS ADAPTIVE COMBINER
C               TO REALIZE ADAPTIVE NOISE CANCELLING FILTER
C
C       INPUT:
C               1. UNFORMATTED INTEGER DATA FILE HOLDING
C                  SIGNAL SAMPLES (PRIMARY INPUT)
C
C               2. UNFORMATTED INTEGER FILE HOLDING REFERENCE
C                  SIGNAL SAMPLES.
C
C       OUTPUT:
C
C               1. UNFORMATTED INTEGER FILE HOLDING FILTERED
C                  OUTPUT SAMPLES.
C
C
C       LINK:   LMS,WFILE,RFILE
C
C
C       REFERENCES:
C
C               1. COHEN,A. BIOMEDICAL SIGNAL PROCESSING
C                  CRC PRESS, CHAPTER 9.
C
C
C               2. WIDROW,B., ET AL, ADAPTIVE NOISE CANCELLING
C                  PRINCIPLES AND APPLICATIONS PROC. IEEE.,63,1692,1975
C
C
C
      DIMENSION ISI(10000),ISR(10000),ISO(10000),IAUX(1024)
      REAL MU,X(50),W(50)
      INTEGER ORD
      BYTE NAMEI(11),NAMER(11),NAMEO(11)
C
C               READ INPUT AND REFERENCE FILES
C
      TYPE 1000
1000  FORMAT(//X'DATA INPUT FILE: ')
      CALL RFILE(NAMEI,ISI,NSAMI,IAUX)
      TYPE 1001
1001  FORMAT(//X'REFERENCE INPUT FILE: ')
      CALL RFILE(NAMER,ISR,NSAMR,IAUX)
C
C               GET FILTER'S PARAMETERS

     * 'ESTIMATION OF PSD BY THE MAX. ENTROPY ALGORITHM (AR)')
      PRINT 111,(NAME(I),I=1,11)
111   FORMAT(/10X'INPUT FILE NAME: '11A1)
      PRINT 112,NTS
112   FORMAT(10X'TOTAL NO. OF SAMPLES: 'I4)
      PRINT 113,NAR
113   FORMAT(10X'ORDER OF AR MODEL: 'I4)
      PRINT 114,(NAME1(I),I=1,11),IT
114   FORMAT(/10X'OUTPUT FILE NAME: '11A1/
     * 10X'NO. OF RECORDS: 1  NO. OF SAMPLES: 'I4)
      STOP
      END
```

```
C
        TYPE 5
5       FORMAT(/H$'GIVE FILTERS PARAMETERS: MU, ORDER AND GAIN: ')
        ACCEPT *,MU,ORD,GAIN
C
C               PREPARE REFERENCE VECTOR
C
        DO 102 I=1,ORD-1
102     X(I)=ISR(ORD-I)*GAIN
C
C               INITIATE WEIGHTING VECTOR & EPSI
C
        DO 103 I=1,ORD
103     W(I)=0.
        EPSI=ISI(ORD-1)*GAIN
C
C               INITIATE OUTPUT VECTOR
C
        DO 106 I=1,ORD
106     ISO(I)=ISI(I)
C*****************************************************
C
C               START FILTERING DATA
C
104     CONTINUE
        DO 105 J=ORD,NSAMI
C
C               UPDATE REFERENCE VECTOR
C
        DO 107 I=ORD,2,-1
107     X(I)=X(I-1)
        X(1)=ISR(J)*GAIN
C
C               GET NEW DESIRED OUTPUT SAMPLE
C
        D=ISI(J)*GAIN
        CALL LMS(X,ORD,W,MU,EPSI,Y,D)
        ISO(J)=INT(EPSI/GAIN+0.5)
105     CONTINUE
C*****************************************************
C
C               WRITE OUTPUT FILE
C
        CALL WFILE(NAMEO,ISO,NSAMI,NORO)
C
C               PROGRAMS DETAILS
C
        TYPE 201
201     FORMAT(//10X' RESULTS OF ADPCAN PROGRAM'/)
        TYPE 200,(NAMEI(I),I=1,11)
200     FORMAT(5X'INPUT FILE NAME: '11A1)
        TYPE 202,(NAMER(I),I=1,11)
202     FORMAT(/5X'REFERENCE FILE NAME: '11A1)
        TYPE 203,(NAMEO(I),I=1,11)
203     FORMAT(/5X'OUTPUT FILE NAME: '11A1)
        STOP
        END
```

```
C
      PROGRAM CONLIM
C                     (VAX VMS VERSION)
C
C
C
C
C     THIS PROGRAM DETECTS WAVLETS BY MEANS OF THE CONTOUR LIMITING
C     METHOD. THE PROGRAM READS THE TEMPLATE SIGNAL (ITEMP) FROM
C     A FILE.
C     UPPER AND LOWER CONTOURS ARE DEFINED:
C
C           UPPER CONTOUR(I)=ITEMP(I)+((EPSI)*ITEMP(I)+CON)
C           LOWER CONTOUR(I)=ITEMP(I)-((EPSI)*ITEMP(I)+CON)
C
C     THE PROGRAM THEN READS THE ACTUAL NOISY WAVLETS SIGNAL,IS(I),
C     FROM ANOTHER FILE AND DETECTS THE PRESENCE OF THE WAVELETS.
C          DETECTION LAW:
C
C          WAVELET IS DETECTED IF FOR AT LEAST 0.9*NOPT SAMPLES
C          (NOPT BEING THE NO. OF SAMPLES IN THE TEMPLATE) WE
C          HAVE:
C
C            LOWER CONTOUR(I).LE.IS(I).LE.UPPER CONTOUR(I)
C
C     INPUT FILES:
C                  1. UNFORMATTED INTEGER TEMPLATE FILE WITH
C                     ONE RECORD AND NOPT SAMPLES.
C                  2. UNFORMATTED INTEGER SIGNAL FILE WITH NREC
C                     RECORDS AND NOPS SAMPLES PER RECORD.
C
C     OUTPUT FILE:
C                  1.UNFORMATTED INTEGER FILE WITH 3 RECORDS
C                    AND NOPT SAMPLES PER RECORD. THE RECORDS:
C                         1.THE TEMPLATE
C                         2.UPPER CONTOUR
C                         3.LOWER CONTOUR
C                  2.UNFORMATTED INTEGER FILE WITH NREC RECORDS,
C                    NOPS SAMPLES PER RECORD. THE FILE CONTAINS
C                    THE LOCATIONS OF DETECTED WAVELETS (DETECTED
C                    WAVELET IS DENOTED BY A PULSE OF AMPLITUDE
C                    OF 512)
C
C
C
C      REFERENCE:

C               COHEN,A. BIOMEDICAL SIGNAL PROCESSING
C                        CRC PRESS, CHAPTER 1
C
C
C
C
C
C
C
C
C
C
      INTEGER ITEMP(2048),IS(4096),IP(4096)
      BYTE NAME1(11),NAME2(11),NAME3(11),NAME4(11)
C
C
C           READ TEMPLATE FILE
C
C
      TYPE 100
100   FORMAT(H$'ENTER INPUT TEMPLATE FILE NAME: ')
      ACCEPT 119,NCH1,(NAME1(I),I=1,11)
119   FORMAT(Q,11A1)
      TYPE 101
```

```
101     FORMAT(H$'ENTER NO. OF SAMPLES IN TEMPLATE: ')
        ACCEPT *,NOPT
        CALL ASSIGN (1,NAME1,11)
        NOPT2=NOPT*2
        DEFINE FILE 1(1,NOPT2,U,IVAR)
        TYPE *,'NO. OF SAMPLES IN TEMP= ',NOPT
        READ(1'1)(ITEMP(I),I=1,NOPT)
        CALL CLOSE(1)
C
C
C               READ SIGNAL'S SAMPLES FROM INPUT FILE
C
C
        TYPE 102
102     FORMAT(H$'ENTER NAME OF SIGNAL FILE: ')
        ACCEPT 119,NCH2,(NAME2(I),I=1,11)
        TYPE 103
103     FORMAT(H$'ENTER NO. OF RECORDS AND SAMPLES PER RECORD: ')
        ACCEPT *,NREC,NOPS
        NOPS2=NOPS*2
C
C
C               PREPARE OUTPUT FILES
C
C
        TYPE 104
104     FORMAT(H$'ENTER CONTOURS OUTPUT FILE NAME: ')
        ACCEPT 119,NCH3,(NAME3(I),I=1,11)
        TYPE 105
105     FORMAT(H$'ENTER SIGNAL OUTPUT FILE NAME: ')
        ACCEPT 119,NCH4,(NAME4(I),I=1,11)
        CALL ASSIGN(2,NAME2,11)

        DEFINE FILE 2(NREC,NOPS2,U,IVAR)
        CALL ASSIGN(4,NAME4,11)
        DEFINE FILE 4(NREC,NOPS2,U,IVAR)
C
C
C
C
        TYPE 106
106     FORMAT(H$'ENTER CONSTANT AND RELATIVE CONTOUR PARAMETERS: ')
        ACCEPT *,CON,EPSI
C
C           LOOP ON ALL SIGNAL RECORDS
C
        NOPTH=INT(NOPT/2+0.5)
        NOPTL=INT(NOPT*0.9+0.5)
        NMODS=NOPS+NOPT-1             !NO. OF SAMPLES IN BUFFER
C
        DO 9 K=1,NREC
C
C              TRANSFER REMAINDER OF PREVIOUS BUFFER TO CURRENT BUFFER
C
C
C                        FIRST RECORD READ DIRECTLY (NO REMENANCE)
        IF (K.LT.1) GO TO 12
        READ(2'1)(IS(J),J=1,NOPS)
        GO TO 13
12      DO 3 J=1,(NOPT-1)
3       IS(1)=IS(NOPS-NOPT+J+1)
        READ(2'K)(IS(J),J=NOPT,NMODS)
13      DO 4 J=1,NOPS
        IP(J)=0
        IC=0
        DO 5 JJ=1,NOPT
        CORF=ITEMP(JJ)*EPSI+CON
        H=ITEMP(JJ)+CORF
        L=ITEMP(JJ)-CORF
        JM1=J+JJ-1
```

```
5       IF(L.LE.IS(JM1).AND.IS(JM1).LE.H) IC=IC+1
        IF(IC.LT.NOPTL)GO TO 4
C
C       A WAVELET HAS BEEN DETECTED
C
        IP(J)=512
4       CONTINUE
        WRITE(4'K)(IP(J),J=1,NOPS)

9       CONTINUE
C
C         END DETECTION
C
        CALL CLOSE(2)
        CALL CLOSE(4)
        CALL ASSIGN(3,NAME3,11)
        DEFINE FILE 3(3,NOPT2,U,IVAR)
        WRITE(3'1) (ITEMP(I),I=1,NOPT)
        DO 10 I=1,NOPT
10      ITEMP(I)=ITEMP(I)*(1.+EPSI)+CON

        WRITE(3'2) (ITEMP(I),I=1,NOPT)
        DO 11 I=1,NOPT
        X=ITEMP(I)
11      ITEMP(I)=((X-CON)/(1.+EPSI))*(1.-EPSI)-CON
        WRITE(3'3) (ITEMP(I),I=1,NOPT)
        STOP
        END


C
C

        PROGRAM COMPRS
C
C                       (VAX VMS VERSION)
C
C
C
C       THIS PROGRAM COMPUTES A TRANSFORMATION MATRIX, TO REDUCE THE
C
C       PATTERN SPACE, BY THREE METHODS: KARHUNEN LOEVE (KL), ENTROPY
C
C       MINIMIZATION (ENT) AND FISHER DISCRIMINANT (FI).
C
C       THE TWO CLASS PROBLEM IS CONSIDERED BY THE PROGRAM.
C
C       PROBABILITIES OF CLASS APPEARENCE HAS BEEN ASSUMED
C
C       EQUAL TO 0.5 FOR THE TWO CLASSES.
C
C       COMPRESSION IS PERFORMED TO PRESERVE SEPERABILITY OF
C
C       CLASSES ACCORDING TO THE VARIOUS CRITERIA OF THE THREE METHODS.
C
C       THE FIRST AND SECOND METHODS REDUCE THE N DIMENSIONAL PATTERN
C
C       SPACE  TO AN M DIMENSIONAL FEATURE SPACE (M < N).
C
C       FISHER DISCRIMINANT REDUCES THE DIMENSIONALITY TO M=1.
C
C
C
C       INPUT:
C                1. A DATA FILE HOLDING A MATRIX OF L1 VECTORS
C                   OF DIMENSION N, CORRESPONDING TO THE MEMBERS
C                   OF THE FIRST CLUSTER.
C
C                2. A DATA FILE HOLDING A MATRIX OF L2 VECTORS
C                   OF DIMENSION N, CORRESPONDING TO THE MEMBERS
C                   OF THE SECOND CLUSTER.
```

```
C
C       OUTPUT:
C               1. A DATA FILE HOLDING M VECTORS OF DIMENSION N,
C                  CORRESPONDING TO THE TRANSFORMATION MATRIX
C                  OF THE COMPRESSION. IN THE CASE OF FISHER

C                  METHOD M=1.
C
C
C
C
C       REFERENCES:
C
C                  1. COHEN,A., BIOMEDICAL SIGNAL PROCESSING,
C                          CRC PRESS, CHAPTER 12
C
C                  2. DUDA,R.,O.,AND HART,P.,E., PATTERN
C                          CLASSIFICATION AND SCENE ANALYSIS, WILEY
C                          INTERSCIENCE, N.Y., 1973
C
C                  3. FUKUNAGA,K., INTRODUCTION TO STATISTICAL
C                          PATTERN RECOGNITION, ACADEMIC PRESS,
C                          N.Y., 1972
C
C                  4. TOU,J.,T., AND GONZALEZ,R.,C., PATTERN
C                          RECOGNITION PRINCIPLES, ADDISON-WESLEY,
C                          READING,Ma.,1974
C
C
C
C       LINKING: EIGEN,RFILEM,WFILEM,MEAN,COVA,ADD,INVER,MUL,SYMINV
C
C
        DIMENSION X1(40,500),X2(40,500),XM(40),XM1(40),XM2(40)
        DIMENSION R2(40,40),R1(40,40),C1(40,40),C2(40,40)
        DIMENSION CINV(40,40),COR(40,40),COV(40,40)
        DIMENSION DELTA(40),DEL(40,40),A(40,40),WR(40),WI(40)
        INTEGER IAUX(80)
        BYTE NAME1(11),NAME2(11)



        TYPE*, '     PLEASE SELECT A REDUCTION METHOD:'
        TYPE*
        TYPE*,'      K-L (FROM N-DIM. TO M DIM.)....... TYPE <1>  AND <CR>'
        TYPE*,'      ENT (FROM N-DIM. TO M DIM.)....... TYPE <2>  AND <CR>'
        TYPE*,'      FIS (FROM N-DIM. TO 1 DIM.)....... TYPE <3>  AND <CR>'
        TYPE*
 803    TYPE*,'      PLEASE TYPE THE METHOD CODE: '
        ACCEPT*,ME
 C
        M=1
        IF (ME.EQ.3) GO TO 807
        TYPE *,'     GIVE DIMENSION OF REDUCED SPACE: '
        ACCEPT *,M
 807    CONTINUE
 C
 C               READ INPUT FILES
 C
 C
        TYPE 777,1
 777    FORMAT(/3X'READ DATA FROM FILE OF CLASS NO.: 'I2/)
        CALL RFILEM(NAME1,X1,40,500,L1,N)

        TYPE 777,2
        CALL RFILEM(NAME2,X2,40,500,L2,N)
 C
 C
 C               MEAN CALCULATIONS
 C
        CALL MEAN(X1,40,500,N,L1,XM1)
```

```
      CALL MEAN(X2,40,500,N,L2,XM2)
C
C     COVARIANCE CALCULATION
C
      CALL COVA(X1,40,500,N,L1,XM1,C1)
      CALL COVA(X2,40,500,N,L2,XM2,C2)
C
C
C     COMMON COVARIANCE
C
      FA=0.5
      CALL ADD(C1,C2,COV,40,40,N,N,1)
      DO 480 I=1,N
      DO 480 J=1,N
480   COV(I,J)=FA*COV(I,J)
C
C.......INVERSE OF COMMON COVARIANCE
C
      CALL INVER(COV,40,40,N,CINV)!CINV:INVERSE OF COMMON COVARIANCE
      N1=1
C
C
C             FISHER METHOD
C
      IF(ME.NE.3) GO TO 800
C
C             PREPARE MEAN DIFFERENCE
C
      CALL ADD(XM1,XM2,DELTA,40,1,N,N1,-1) !DELTA IS
C                                  THE DIFFERENCE IN CLUSTERS MEANS.
C
C             CALCULATE FISHER VECTOR (CINV*DELTA)
C
      CALL MUL(CINV,40,40,DELTA,40,1,DEL,40,1,N,N,N1)
C
C             NORMALIZE FISHER VECTOR
C
      XXN=0.
      DO 810 J=1,N
810   XXN=XXN+DEL(J,1)*DEL(J,1)
      XXN=SQRT(XXN)
      DO 811 J=1,N
811   A(J,1)=DEL(J,1)/XXN   !FIRST ROW OF A HOLDS NORM. FISHER
      GO TO 778
800   CONTINUE
C
C             MINIMUM ENTROPY METHOD
C
      IF(ME.NE.2) GO TO 801
      CALL EIGEN(40,N,COV,WR,WI,A,IERR,WO)

      GO TO 778
801   CONTINUE
C
C             K-L METHOD
C
      IF(ME.NE.1)GOTO 77
C
C     COMMON CORRELATION
C
      DO 600 I=1,N
600   XM(I)=0.                      !XM IS A NULL VECTOR DUMMY MEAN
      CALL COVA(X1,40,500,N,L1,XM,R1) !R1 IS CLUSTER 1 CORRELATION
      CALL COVA(X2,40,500,N,L2,XM,R2) !R2 IS CLUSTER 2 CORRELATION
      CALL ADD(R1,R2,COR,40,40,N,N,1)!COR IS THE COMMON CORRELATION
      DO 451 I=1,N
      DO 451 J=1,N
      COR(I,J)=COR(I,J)/2.
451   CONTINUE
      CALL EIGEN(40,N,COR,WR,WI,A,IERR,WO)
C
```

```
C               CHANGE ORDER OF ROWS TO GET EIGENVECTORS
C               CORRESPONDING TO M LARGEST EIGENVALUES
C               (R1 IS DESTROYED)
C
        DO 808 I=1,N
        DO 808 J=1,M
        JJ=N-J+1
808     R1(I,J)=A(I,JJ)
        DO 809 I=1,N
        DO 809 J=1,M
809     A(I,J)=R1(I,J)
        GO TO 778
C
77      TYPE*,' SORRY, SELECT A METHOD AGAIN (CODE 1-3)!!!)'
        GO TO 803
778     CONTINUE
C
C               WRITE TRANSFORMATION ON OUTPUT FILE
C
        CALL WFILEM(NAME1,A,40,40,M,N)
        END
```

III. SUBROUTINES

```
        SUBROUTINE LMS(X,N,W,MU,EPSI,Y,D)
C
C
C               THIS SUBROUTINE REALIZES THE ADAPTIVE LINEAR
C               COMBINER BY MEANS OF WIDROW'S ALGORITHM
C
C
C               X-   THE REFERENCE VECTOR
C               N-   THE ORDER OF X
C               W-   THE WEIGHTING VECTOR
C                        SUBROUTINE RECIEVES CURRENT WEIGHTES
C                        AND TRANSMITS PREDICTED WEIGHTES
C              MU-   THE GRADIENT CONSTANT
C            EPSI-   CURRENT OUTPUT,
C                        FILTERED SIGNAL
C               Y-   ADAPTIVE LINEAR COMBINER'S OUTPUT,
C                        ESTIMATION OF NOISE
C
C               D-   PRIMARY INPUT SAMPLE
C
C       REFERENCE:
C
C               WIDROW,B. ET AL, ADAPTIVE NOISE CANCELLING
C                    PRINCIPLES AND APPLICATIONS,PROC. IEEE,
C                    63,1692,1975
C
C
        DIMENSION W(1),X(1)
        REAL MU,EPSI,Y,D
C
C
C               CALCULATE CURRENT COMBINER'S OUTPUT
C
        Y=0.0
        DO 2 I=1,N
2       Y=Y+W(I)*X(I)
C
C               CALCULATE CURRENT OUTPUT
C
        EPSI=D-Y
C
```

```
C             UPDATE WEIGHTING VECTOR
C
        CONS=2*MU*EPSI
        DO 1 I=1,N
        W(I)=W(I)+CONS*X(I)
1       CONTINUE
        RETURN
        END




        SUBROUTINE NACOR(S,L,COR,NP1,ENG)
C
C
C
C       THIS SUBROUTINE COMPUTES THE NORMALIZED AUTOCORRELATION
C       SEQUENCE (IN NORMAL USE, AS AN INPUT TO 'DLPC',
C       THE L.P.C EXTRACTION SUBROUTINE).
C
C
C
C
C       S.......................INPUT VECTOR
C       L.......................THE DIMENSION OF 'S'
C       COR.....................VECTOR OF NORMALIZED AUTOCORRELATION
C       NP1.....................THE DIMENSION OF 'COR'+1
C                               (CORRELATION COEFF. FROM ZERO TO N)
C       ENG.....................THE ENERGY OF THE INPUT.
C
C
C
C
C
        REAL S(1),COR(1)
        ENG=0
        DO 5 I=1,L
5       ENG=ENG+S(I)**2
        DO 10 I=2,NP1
        COR(I)=0
        DO 20 J=1,L+1-I
20      COR(I)=COR(I)+S(J)*S(J+I-1)
10      COR(I)=COR(I)/ENG
        COR(1)=1
        RETURN
        END
```

```
      SUBROUTINE DLPC(P,COR,LPC,PAR,AUX,ERR)
C
C
C
C
C       THIS SUBROUTINE COMPUTES THE LPC,
C       THE PARCOR COEF. AND THE TOTAL SQUARED ERROR
C       OF A SEQUENCE, OUT OF THE AUTOCORRELATION SEQUENCE.
C
C
C     DESCRIPTION OF PARAMETERS
C
C     P.......................DIMENSION OF 'LPC','PAR' & 'AUX'.
C     COR.....................#P+1 AUTO-CORR. COEF. VECTOR.
C     LPC.....................LPC COEF. VECTOR.
C     PAR.....................PARCOR COEF VECTOR.
C     AUX.....................WORKING AREA.
C     ERR.....................NORMALIZED PREDICTION ERROR.
C
C
C     REFERENCE:
C
C                 1. MAKHOUL,J.,' LINEAR PREDICTION:A TUTORIAL
C                       REVIEW',PROC. IEEE,63,561,1975
C
C
C     LINK: NONE
C
C
C
      REAL COR(1),LPC(1),PAR(1),AUX(1)
      INTEGER P
      PAR(1)=-COR(2)
      LPC(1)=PAR(1)
      ERR=(1-PAR(1)**2)
      DO 10 I=2,P
      I1=I-1
      TEMP=-COR(I+1)
      DO 20 J=1,I1
20    TEMP=TEMP-LPC(J)*COR(1+I-J)
      PAR(I)=TEMP/ERR
      LPC(I)=PAR(I)
      DO 30 K=1,I1
30    AUX(K)=LPC(K)+PAR(I)*LPC(I-K)
      DO 40 L=1,I1
40    LPC(L)=AUX(L)

10    ERR=ERR*(1-PAR(I)**2)
      RETURN
      END
```

```
      SUBROUTINE DLPC20(COR,XMAT)
C
C
C
C
C     THIS SUBROUTINE COMPUTES THE LPC
C     AND THE TOTAL SQUARED ERROR FOR
C     ALL THE PREDICTORS FROM ORDER 1 UP TO 20.
C
C
C     DETAILS AND COMMENTS SEE SUBROUTINE DLPC
C
C
C
C     COR..................21 AUTO-CORR. COEF. VECTOR
C     MAT..................MATRIX WHICH CONTAINS THE SOLUTION OF ALL
C                          THE PREDICTORS.THE FIRST COLUMN CONTAINS
C                          THE NORMALIZED ERROR.
C
      REAL COR(21),XMAT(20,21),PAR(20),LPC(20),Y(25)
      PAR(1)=-COR(2)
      LPC(1)=PAR(1)
      ERR=(1-PAR(1)**2)
      XMAT(1,1)=ERR
      XMAT(1,2)=LPC(1)
      DO 60 I=2,20
      I1=I-1
      TEMP=-COR(I+1)
      DO 20 J=1,I1
20    TEMP=TEMP-LPC(J)*COR(1+I-J)
      PAR(I)=TEMP/ERR
      LPC(I)=PAR(I)
      DO 30 K=1,I1
30    Y(K)=LPC(K)+PAR(I)*LPC(I-K)
      DO 40 L=1,I1
40    LPC(L)=Y(L)
      ERR=ERR*(1-PAR(I)** 2)
      XMAT(I,1)=ERR
      DO 60 J=1,20
      XMAT(I,J+1)=LPC(J)
60    CONTINUE
C     DO 70,I=1,20
C     TYPE*,I
C     TYPE*,XMAT(I,1),'""""""""',XMAT(I,2)
C70   CONTINUE

      RETURN
      END
```

```
      SUBROUTINE FT01A (IT,INV,TR,TI)
C
C
C         THIS ROUTINE CALCULATES THE DISCRETE FOURIER TRANSFORM OF
C         THE SEQUENCE F(N);  N<0,1,...,IT-1
C         THE DATA IS TAKEN TO BE PERIODIC NAMELY:  F(N+IT) = F(N).
C         IT IS THE SEQUENCE DIMENSION AND MUST BE A POWER OF 2.
C         THE PROGRAM CALCULATES THE DIRECT TRANSFORM (INV=2),
C         FOR WHICH:
C
C         G(M) = SUM OVER N=0,1,..,IT-1 OF F(N)*EXP(2PI*SQRT(-1)*N*M/IT)
C         FOR M=0,1,...,IT-1
C
C         IT ALSO CALCULATES THE INVERSE TRANSFORM (INV=1), FOR WHICH:
C
C         F(N) = (1./IT)*(SUM OVER M=0,1,..,IT-1 OF G(M)*EXP(-2PI*SQRT(-1)*
C         N*M/IT) FOR N =0,1,...,IT-1
C
C         IF IT IS NOT A POWER OF 2, INV IS SET TO -1 FOR ERROR RETURN.
C
C         THE SUBROUTINE ACCEPTS REAL AND IMAGINARY PARTS OF SEQUENCE
C         TO BE TRANSFORMED IN ARRAYS TR (REAL) AND TI (IMAGINARY).
C         TRANSFORMED RESULT IS RETURNED IN THESE TWO ARRAYS.
C
C         NOTE THAT WHEN DIRECT FFT OF A REAL SIGNAL IS PERFORMED
C         SIGNAL SHOULD BE IN ARRAY TR AND ARRAY TI MUST BE ZEROED.
C
C
C
C       REFERENCE:
C
C                      1.  GENTLEMAN AND SANDE,
C                          PROC. FALL JOINT COMPUTER CONFER. 1966
C
C
C
C       IT................. SEQUENCE DIMENSION(MUST BE POWER OF 2 !!).
C       IV................. 1 -- INVERSE TRANSFORM.
C                           2 -- DIRECT TRANSFORM.
C       TR................. REAL PART ARRAY.
C       TI................. IMAGINARY PART ARRAY.
C
C
C
      DIMENSION TR(4),TI(4),UR(15),UI(15)
      INTEGER KJUMP
      KJUMP=1

      GO TO(100,200),KJUMP
  100 UM=.5
      DO 50 I=1,15
      UM=.5*UM
      TH=6.283185307178*UM
      UR(I)=COS(TH)
   50 UI(I)=SIN(TH)
  200 UM=1.
      GO TO(1,2),INV
    1 UM=-1.
    2 IO=2
      DO 3 I=2,16
      IO=IO+IO
      IF(IO-IT)3,4,5
    3 CONTINUE
C     ERROR IN IT - SET INV=-1 AND RETURN
    5 INV=-1
      RETURN
C     IT= 2**I - INITIALISE OUTER LOOP
    4 IO=I
      II=IO
      I1=IT/2
      I3=1
```

```
C       START MIDDLE LOOP
   10 K=0
      I2=I1+I1
C     CALCULATE TWIDDLE FACTOR E(K/I2)
   11 WR=1.
      WI=0.
      KK=K
      J0=I0
   24 IF(KK)21,22,21
   21 J0=J0-1
      KK1=KK
      KK=KK/2
      IF(KK1-2*KK)23,21,23
   23 WS=WR*UR(J0)-WI*UI(J0)
      WI=WR*UI(J0)+WI*UR(J0)
      WR=WS
      GO TO 24
   22 WI=WI*UM
C     START INNER LOOP
      J=0
C     DO 2*2 TRANSFORM
   31 L=J*I2+K
      L1=L+I1
      ZR=TR(L+1)+TR(L1+1)
      ZI=TI(L+1)+TI(L1+1)
      Z=WR*(TR(L+1)-TR(L1+1))-WI*(TI(L+1)-TI(L1+1))
      TI(L1+1)=WR*(TI(L+1)-TI(L1+1))+WI*(TR(L+1)-TR(L1+1))
      TR(L+1)=ZR
      TR(L1+1)=Z
      TI(L+1)=ZI
C     INDEX J LOOP
      J=J+1
      IF(J-I3)31,12,12
C     INDEX K LOOP

   12 K=K+1
      IF(K-I1)11,6,6
C     INDEX OUTER LOOP
    6 I3=I3+I3
      I0=I0-1
      I1=I1/2
      IF(I1)51,51,10
C     UNSCRAMBLE
   51 J=1
      UM=1.
      GO TO(61,52),INV
   61 UM=1./FLOAT(IT)
   52 K=0
      J1=J
      DO 53 I=1,II
      J2=J1/2
      K=2*(K-J2)+J1
   53 J1=J2
   54 IF(K-J)66,56,55
   56 TR(J+1)=TR(J+1)*UM
      TI(J+1)=TI(J+1)*UM
      GO TO 66
   55 ZR=TR(J+1)
      ZI=TI(J+1)
      TR(J+1)=TR(K+1)*UM
      TI(J+1)=TI(K+1)*UM
      TR(K+1)=ZR*UM
      TI(K+1)=ZI*UM
   66 J=J+1
      IF(J-IT+1)52,57,57
   57 TR(1)=TR(1)*UM
      TI(1)=TI(1)*UM
      TR(IT)=TR(IT)*UM
      TI(IT)=TI(IT)*UM
      RETURN
      END
```

```
      SUBROUTINE XTERM(X,N,XMAX,XMIN)
C
C
C
C           THIS SUBROUTINE FINDS THE MAXIMAL AND MINIMAL
C           VALUES OF A GIVEN (REAL) VECTOR
C
C
C                  X----------- REAL VECTOR
C                  N----------- DIMENSIONS OF X
C               XMAX----------- MAXIMAL VALUE OF X
C               XMIN----------- MINIMAL VALUE OF X
C
C
      REAL X(1)
      XMIN=99E+35
      XMAX=-XMIN
      DO 1 I=1,N
      IF (X(I).GT.XMAX) XMAX=X(I)
1     IF (X(I).LT.XMIN) XMIN=X(I)
      RETURN
      END
```

```
      SUBROUTINE ADD(A,B,F,N1,N2,N,M,NCA)
C
C              SUB FOR MATRIX ADITION OR SUBSTRACTION
C
C              A,B,F.........A AND B ARE THE MATRICES TO
C                            BE ADDED OR SUBTRACTED, THE
C                            RESULT IS STORED IN F.
C
C              N1,N2.........NO. OF ROWS AND COLUMNS OF
C                            MATRICES A, B, AND F AS
C                            DIMENSIONED IN MAIN PROGRAM.
C
C              N,M...........DIMENSIONS OF THE DATA IN MATRICES
C
C              NCA...........CA=1    F=A+B
C                            CA=-1   F=A-B
C
C
      DIMENSION A(N1,N2),B(N1,N2),F(N1,N2)
      DO 101 I=1,N
      DO 101 J=1,M
      F(I,J)=A(I,J)+(NCA*B(I,J))
101   CONTINUE
      RETURN
      END
```

```
      SUBROUTINE MUL(A,NA1,NA2,B,NB1,NB2,D,ND1,ND2,N1,N2,N3)
C
C     SUB FOR MATRIX MULTIPLICATION.'A' & 'B' ARE THE MATRICES TO
C     BE MULTIPLIED.'D' IS THE PRODUCT.
C
C             NA1,NA2........DIMENSIONS OF A IN MAIN PROGRAM
C
C             NB1,NB2........DIMENSIONS OF B IN MAIN PROGRAM
C
C             ND1,ND2........DIMENSIONS OF D IN MAIN PROGRAM
C
C             N1,N2..........NO. OF ROWS AND COLUMNS OF A
C                            WITH DATA TO BE MULTIPLIED
C
C             N2,N3..........NO. OF ROWS AND COLUMNS OF B
C                            WITH DATA TO BE MULTIPLIED
C
C             N1,N3..........PART OF D IN WHICH RESULT IS STORED
C
C
      DIMENSION A(NA1,NA2),B(NB1,NB2),D(ND1,ND2)
      DO 350 I=1,N1
      DO 350 K=1,N3
      D(I,K)=0
      DO 40 M=1,N2
40    D(I,K)=D(I,K)+A(I,M)*B(M,K)
350   CONTINUE
      RETURN
      END



C
      SUBROUTINE MEAN(X,N1,N2,N,L,XM)
C
C
C             THIS SUBROUTINE ESTIMATES THE MEAN OF A SET OF
C             VECTORS
C
C
C             X.........A MATRIX OF DIMENSION N*L WHOSE
C                       COLOUMNS ARE THE L, N DIMENSIONAL VECTORS
C                       THE MEAN OF WHICH IS REQUERED.
C
C            N1.........THE NO. OF ROWS OF X AS DIMENSIONED IN
C                       THE MAIN PROGRAM
C
C            N2.........THE NO. OF COLUMNS OF X AS DIMENSIONED
C                       IN THE MAIN PROGRAM
C
C             N.........THE DIMENSION OF THE VECTORS
C
C             L.........THE NUMBER OF VECTORS TO BE AVERAGED
C
C            XM.........THE N DIMENSIONAL MEAN VECTOR
C
      DIMENSION X(N1,N2),XM(N1)
      REAL X,XM
      DO 79 J=1,N
      DO 779 I=1,L
779   XM(J)=XM(J)+X(J,I)
79    XM(J)=XM(J)/L
      RETURN
      END
```

```
      SUBROUTINE COVA(X,N1,N2,N,L,XM,C)
C
C
C               THIS SUBROUTINE ESTIMATES THE COVARINCE MATRIX
C               OF A CLUSTER OF VECTORS
C
C
C               X...........A MATRIX HOLDING L, N DIMENSIONAL VECTORS
C                           DIMENSIONED N1*N2 IN MAIN PROGRAM
C
C              N1...........NO. OF ROWS OF X AS DIMENSIONED IN
C                           MAIN PROGRAM
C
C              N2...........NO. OF COLUMNS OF X AS DIMENSIONED
C                           IN MAIN PROGRAM
C
C               N...........DIMENSION OF THE VECTORS
C
C               L...........NUMBER OF VECTORS
C
C              XM...........AN N DIMENSIONAL MEAN VECTOR
C
C               C...........THE N*N COVARIANCE MATRIX ESTIMATED
C                           BY THE SUBROUTINE. C MUST BE DIMENSIONED
C                           IN MAIN PROGRAM AS N1*N1
C
C
C
C
      DIMENSION X(N1,N2),XM(N1),C(N1,N1)
      REAL X,XM,C
      DO 5 I=1,N
      DO 5 J=1,N
5     C(I,J)=0.
      DO 2 K=1,L
      DO 2 J=1,N
      DO 2 I=1,N
2     C(I,J)=C(I,J)+(X(I,K)-XM(I))*(X(J,K)-XM(J))
      DO 3 J=1,N
      DO 3 I=1,N
3     C(I,J)=C(I,J)/L
      RETURN
      END
```

```
      SUBROUTINE INVER(CMAT,N1,N2,IROW,CMAT1)
C
C           THIS SUBROUTINE COMPUTES THE INVERSE OF A
C           REAL MATRIX WITHOUT DESTROYING THE ORIGINAL ONE
C
C           CMAT.........REAL MATRIX TO BE INVERTED
C
C           N1,N2........DIMENSION OF MATRIX AS DIMENSIONED IN
C                        MAIN PROGRAM
C
C           IROW.........NO. OF ROWS AND COLUMNS OF CMAT
C                        DEFINING THE PART TO BE INVERTED
C
C           CMAT1........THE MATRIX (N1*N1) IN WHICH THE
C                        THE COMPUTED INVERSE IS STORED
C
C
      DIMENSION CMAT(N1,N2),CMAT1(N1,N2)
      DO 500 J=1,IROW
      DO 500 I=1,IROW
500   CMAT1(I,J)=CMAT(I,J)
      CALL SYMINV(CMAT1,N1,N2,IROW,IER)
      IF(IER.NE.1)GOTO 502
      TYPE*,'NO INVERSE FOUND'
      DO 501 J=1,IROW
      DO 501 I=1,IROW
501   CMAT1(I,J)=1.
      GOTO 504
502   TYPE*,'INVERSE FOUND'
504   RETURN
      END
```

```
      SUBROUTINE SYMINV(A,L,M,N,IFAIL)
C
C
C           THIS SUBROUTINE COMPUTES THE INVERSE OF A REAL
C           SYMETRIC PART (N*N) OF A MATRIX. THE GAUSS JORDAN
C           METHOD WITH NORMALISATION BUT WITH NO INTERCHANGE OF
C           ROWS AND COLUMNS IS USED.
C           THE ORIGINAL MATRIX IS NOT SAVED.
C           THE SUBROUTINE IS ADAPTED FROM CERN.
C
C
C           A............A TWO DIMENSIONAL MATRIX
C
C           L............THE NUMBER OF ROWS OF A
C
C           M............THE NUMBER OF COLUMNS OF A
C
C           N............THE NUMBER OF ROWS (AND COLUMNS)
C                        OF THE PORTION OF A TO BE INVERTED
C
C       IFAIL............ERROR FLAG:
C                                  0=INVERSE HAS BEEN FOUND
C                                  1=ALL ELEMENTS ON DIAGONAL
C                                    ARE ZERO, OR THE SYSTEM
C                                    APPEARS TO BE DEPENDENT
C                                    NO INVERSE FOUND.
C
C
C     REFERENCES:
C                1. KOU,S.S., NUMERICAL METHODS AND COMPUTERS,
C
C
      DIMENSION A(L,M),P(100),Q(100),R(100)
      INTEGER R
      IFAIL=0
```

```
      EPSILN=1.E-6
      IF(L.LT.N) GO TO 95
      IF(M.LT.N) GO TO 96
C
C                     CONSTRUCT TRUTH TABLE
C
      DO 10 I=1,N
   10 R(I)=1
C
C                     BEGIN PROGRAMME
C
      DO 65 I=1,N
      K=0
C
C                     SEARCH FOR PIVOT
C
      BIG=0.
      DO 37 J=1,N
      TEST=ABS(A(J,J))

      IF(TEST-BIG)37,37,31
   31 IF(R(J))100,37,32
   32 BIG=TEST
      K=J
   37 CONTINUE
C
C                     TEST FOR ZERO MATRIX
C
      IF(K)100,100,38
C
C                     TEST FOR LINEARITY
C
   38 IF(I.EQ.1) PIVOT1=A(K,K)
      IF(ABS(A(K,K)/PIVOT1)-EPSILN) 100,39,39
C
C                     PREPARATION FOR ELIMINATION STEP1
C
   39 R(K)=0
      Q(K)=1./A(K,K)
      P(K)=1.
      A(K,K)=0.0
      KP1=K+1
      KM1=K-1
      IF(KM1)100,50,40
   40 DO 49 J=1,KM1
      P(J)=A(J,K)
      Q(J)=A(J,K)*Q(K)
      IF(R(J))100,49,42
   42 Q(J)=-Q(J)
   49 A(J,K)=0.
   50 IF(K-N)51,60,100
   51 DO 59 J=KP1,N
      P(J)=A(K,J)
      Q(J)=-A(K,J)*Q(K)
      IF(R(J))100,52,59
   52 P(J)=-P(J)
   59 A(K,J)=0.0
C
C                     ELIMINATION PROPER
C
   60 DO 65 J=1,N
      DO 65 K=J,N
   65 A(J,K)=A(J,K)+P(J)*Q(K)
C
C                     ELEMENTS OF LEFT DIAGONAL
C
      DO 70 J=2,N
      JM1=J-1
      DO 70 K=1,JM1
   70 A(J,K)=A(K,J)
      RETURN
```

```
C                       FAILURE RETURN
C
   95 PRINT 150,L,N
      GO TO 100
   96 PRINT 151,M,N

  100 IFAIL=1
      RETURN
  150 FORMAT(4H1L =,I5,4H N =,I5,33H L SHOULD BE LARGER OR EQUAL TO N)
  151 FORMAT(4H1M =,I5,4H N =,I5,33H M SHOULD BE LARGER OR EQUAL TO N)
      END
```

```
        SUBROUTINE RFILE(NAME,ISAMP,NTS,IAUX)
C
C                           (VAX VMS VERSION)
C
C
C
C
C                         THIS SUBROUTINE READS A DATA VECTOR
C                         FROM AN UNFORMATTED INTEGER DATA FILE.
C
C
C                NAME- A BYTE ARRAY HOLDING THE NAME OF THE FILE
C
C               ISAMP- INTEGER VECTOR WITH DIMENSIONS CORRESPONDING
C                      TO THE TOTAL NO. OF SAMPLES TO BE READ
C
C                IAUX- AN AUXILIARY INTEGER VECTOR WITH DIMENSION
C                      CORRESPONDING TO THE NO. OF SAMPLES IN A RECORD
C
C                 NTS- NO. OF SAMPLES READ INTO THE DATA VECTOR
C
C
C
        INTEGER ISAMP(1),IAUX(1)
        REAL AUX(1)
        BYTE NAME(1)
        TYPE 100
100     FORMAT(H$'TYPE INPUT FILE NAME: ')
        ACCEPT 119,NCHO,(NAME(I),I=1,11)
119     FORMAT(Q,11A1)
        CALL ASSIGN(1,NAME,11)
        TYPE 101
101     FORMAT(1H$,'GIVE NO. OF RECORDS & NO. OF SAMPLES/RECORD: ')
        ACCEPT *,NOR,NOSR
        TYPE 104
104     FORMAT(1H$'GIVE NO. OF INITIAL RECORD & TOTAL NO. OF SAMPLES:')
        ACCEPT *,NIR,NTS
        NOSR2=NOSR*2
        DEFINE FILE 1(NOR,NOSR2,U,IVAR)
        NFR=NTS/NOSR+NIR-1
        IF (NFR.LT.NIR) NFR=NIR
        DO 1 I=NIR,NFR
        II=(I-NIR)*NOSR
1       READ(1'I) (ISAMP(II+J),J=1,NOSR)
C
C       READING THE REMAINING SAMPLES FROM LAST RECORD
C
        NLAST=(NFR-NIR+1)*NOSR
        NREM=NTS-NLAST

        IF (NREM.LT.1) GO TO 3     ! ALL DATA READ NO REMENANCE IN
C                                    NEXT RECORD
        NFR1=NFR+1
        READ(1'NFR1)(AUX(I),I=1,NOSR)
```

```
      DO 2 I=1,NREM
2     ISAMP(NLAST+I)=AUX(I)
3     CONTINUE
      CALL CLOSE (1)
      RETURN
      END
```

```
      SUBROUTINE WFILE(NAME,ISAMP,NTS,NOR)
C
C                       (VAX VMS VERSION)
C
C
C
C
C                      THIS SUBROUTINE WRITES A DATA VECTOR
C                      ON AN UNFORMATTED INTEGER DATA FILE.
C
C
C               NAME- A BYTE ARRAY HOLDING THE NAME OF THE FILE
C
C              ISAMP- INTEGER VECTOR WITH DIMENSIONS CORRESPONDING
C                     TO THE TOTAL NO. OF SAMPLES TO BE WRITTEN
C
C                NOR- NO. OF RECORDS (CALCULATED BY SUBROUTINE)
C
C                NTS- NO. OF SAMPLES WRITTEN ON THE FILE
C
C
C
      INTEGER ISAMP(1)
      REAL AUX(1)
      BYTE NAME(1)

      TYPE 100
100   FORMAT(1H$'GIVE OUTPUT FILE NAME: ')
      ACCEPT 119,NCHO,(NAME(I),I=1,11)
119   FORMAT(Q,11A1)
      CALL ASSIGN(1,NAME,11)
      TYPE 101
101   FORMAT(1H$'GIVE NO. OF SAMPLES/RECORD: ')
      ACCEPT *,NOSR
      NOSR2=NOSR*2
C
C               CALCULATE NO. OF RECORDS
C
      NOR=NTS/NOSR
      IF (NOR.EQ.0) NOR=1
      NLAST=NOR*NOSR
      NREM=NTS-NLAST
      DEFINE FILE 1(NOR,NOSR2,U,IVAR)
      DO 1 I=1,NOR
      II=(I-1)*NOSR
1     WRITE(1'I) (ISAMP(II+J),J=1,NOSR)
      IF (NREM.LT.1) GO TO 3
      NOR=NOR+1
      WRITE(1'NOR)(ISAMP(J+NLAST),J=1,NREM)

3     CONTINUE
      CALL CLOSE (1)
      RETURN
      END
```

```
      SUBROUTINE RFILEM(NAME,A,N1,N2,NOR,NOSR)
C
C                          (VAX VMS VERSION)
C
C
C                        THIS SUBROUTINE READS A DATA MATRIX
C                        FROM AN UNFORMATTED INTEGER DATA FILE.
C
C                 NAME- A BYTE ARRAY HOLDING THE NAME OF THE FILE
C
C                    A- REAL MATRIX WITH DIMENSIONS N1*N2, SO
C                       DIMENSIONED IN THE MAIN PROGRAM
C
C                   N1- NO. OF ROWS OF MATRIX A, AS DIMENSIONED
C                       IN THE MAIN PROGRAM
C
C                   N2- NO. OF COLUMNS OF MATRIX A, AS DIMENSIONED
C                       IN MAIN PROGRAM
C
C                  NOR- NUMBER OF ROWS IN THE MATRIX, THE NUMBER OF
C                        RECORDS IN THE FILE
C
C                 NOSR- NUMBER OF COLUMNS IN THE MATRIX EQUALS THE
C                       NUMBER OF SAMPLES/RECORD
C
      REAL A(N1,N2)
      BYTE NAME(1)
      TYPE 100
100   FORMAT(H$'TYPE INPUT FILE NAME: ')
      ACCEPT 119,NCHO,(NAME(I),I=1,11)
119   FORMAT(Q,11A1)
      CALL ASSIGN(1,NAME,11)
      TYPE 101
101   FORMAT(1H$,'GIVE NO. OF RECORDS & NO. OF SAMPLES/RECORD: ')
      ACCEPT *,NOR,NOSR
      TYPE 104
104   FORMAT(1H$'GIVE INITIAL AND FINAL RECORDS TO READ: ')
      ACCEPT *,NIR,NFR
      NOSR2=NOSR*2
      DEFINE FILE 1(NOR,NOSR2,U,IVAR)
      IF (NFR.LT.NIR) NFR=NIR
      DO 1 I=NIR,NFR
      II=I-NIR+1
1     READ(1'II) (A(J,II),J=1,NOSR)
      CALL CLOSE (1)
      RETURN
      END
```

```
      SUBROUTINE WFILEM(NAME,A,N1,N2,NOR,NOSR)
C
C                         (VAX VMS VERSION)
C
C
C
C                      THIS SUBROUTINE WRITES A  REAL DATA MATRIX
C                      OF DIMENSIONON NOR*NOSR ON AN UNFORMATTED
C                      REAL DATA FILE.
C
C
C               NAME- A BYTE ARRAY HOLDING THE NAME OF THE FILE
C
C                  A- A REAL MATRIX OF DIMENSION N1*N2 SO DIMENSIONED
C                     IN THE MAIN PROGRAM
C
C                 N1- NO. OF ROWS OF THE MATRIX AS DIMENSIONED
C                     IN MAIN PROGRAM
C
C                 N2- NO. OF COLUMNS OF THE MATRIX AS DIMENSIONED
C                     IN MAIN PROGRAM
C
C                NOR- NO. OF RECORDS IN FILE
C
C               NOSR- NO. OF SAMPLES/RECORD
C
C
      DIMENSION A(N1,N2)
      BYTE NAME(1)
      TYPE 100
100   FORMAT(1H$'GIVE OUTPUT FILE NAME: ')
      ACCEPT 119,NCHO,(NAME(I),I=1,11)
119   FORMAT(Q,11A1)
      CALL ASSIGN(1,NAME,11)
      TYPE 101
101   FORMAT(1H$'GIVE NO. OF RECORDS AND SAMPLES/RECORD: ')
      ACCEPT *,NOR,NOSR
      NOSR2=NOSR*2
      DEFINE FILE 1(NOR,NOSR2,U,IVAR)
      DO 1 J=1,NOR
      TYPE *,(A(I,J),I=1,NOSR)
1     WRITE(1'J) (A(I,J),I=1,NOSR)
3     CONTINUE
      CALL CLOSE (1)
      RETURN
      END
```

INDEX

A

B

C

D

E

P